Innovation Diffusion Models

Innovation Diffusion Models

Theory and Practice

Mariangela Guidolin
University of Padua
Padua, Italy

Registered Offices
John Wiley & Sons, Inc., 111 River Street, Hoboken, NJ 07030, USA
John Wiley & Sons Ltd, The Atrium, Southern Gate, Chichester, West Sussex, PO19 8SQ, UK

For details of our global editorial offices, customer services, and more information about Wiley products visit us at www.wiley.com.

Wiley also publishes its books in a variety of electronic formats and by print-on-demand. Some content that appears in standard print versions of this book may not be available in other formats.

Library of Congress Cataloging-in-Publication Data:

Names: Guidolin, Mariangela, author.
Title: Innovation diffusion models : theory and practice / Mariangela
    Guidolin.
Description: Chichester, West Sussex, UK : Wiley, 2024. | Includes
    bibliographical references and index.
Identifiers: LCCN 2023022913 (print) | LCCN 2023022914 (ebook) | ISBN
    9781119756200 (cloth) | ISBN 9781119756217 (adobe pdf) | ISBN
    9781119756224 (epub)
Subjects: LCSH: New products–Marketing–Mathematical models. | Diffusion
    of innovations.
Classification: LCC HF5415.153 .G88 2024 (print) | LCC HF5415.153 (ebook)
    | DDC 658.5/750688–dc23/eng/20230717
LC record available at https://lccn.loc.gov/2023022913
LC ebook record available at https://lccn.loc.gov/2023022914

Cover design: Wiley
Cover image: © AF-studio/DigitalVision Vectors/Getty Images

Set in 9.5/12.5pt STIXTwoText by Straive, Chennai, India
Printed and bound by CPI Group (UK) Ltd, Croydon, CR0 4YY

C9781119756200_200923

To my family

Contents

List of Figures

List of Tables

About the Author

Mariangela Guidolin is an Associate Professor of Business and Economic Statistics at the Department of Statistical Sciences, University of Padua (Italy). She gained her PhD in economics and management from the University of Padua in 2008 with the thesis "Aggregate and Agent-Based Models for the Diffusion of Innovations."

She currently teaches business statistics courses at the University of Padua in the Master's in Statistical Sciences and Master's in Data Science, where, among other topics, she treats diffusion models and statistical tools for analysis of market penetration. She has a long experience in teaching at different levels (undergraduate, graduate, and extra-university), theory, and applications of statistical analysis of diffusion processes.

Her research interests include innovation diffusion modeling, technological forecasting, and business analytics. She published several research articles about innovation diffusion modeling, with a special focus on the ICT and energy sectors, in journals like *International Journal of Forecasting, Technological Forecasting and Social Change, Renewable and Sustainable Energy Reviews, Physica A,* and *Applied Stochastic Models in Business and Industry.*

Preface

I have always been fascinated by technological evolution and social change occurring through new ideas that spread among individuals. The possibility to model and understand these processes has been at the center of my research interests since my PhD.

In these years, I have had the opportunity to teach innovation diffusion modeling, at undergraduate and graduate levels, to many students with different backgrounds (statistics, mathematics, physics, engineering, economics, and management), and these experiences, with the materials prepared for my lectures, have been a valuable starting point for this book, thanks to the many suggestions and observations received from them.

The book is an introduction to a class of innovation diffusion models used to study the life cycle of new products and technologies. Starting from the basic structure of the Bass model, which gave rise to an extremely productive literature on innovation after its publication in 1969 in *Management Science*, the book focuses on some of its possible generalizations.

Two main generalizing criteria are illustrated: the inclusion of intervention variables and the introduction of dynamic market potential. This has allowed the development of more general models able to capture patterns in data for which the Bass model proves unsuitable. A further direction of generalization is proposed by describing some models for competitive settings. Model building under reasonable assumptions, parsimony, and interpretability is the common thread of all the models discussed throughout the chapters. The main models described in the chapters are always accompanied by real-data applications. An entire chapter is dedicated to the discussion of case studies based on real data, where all the models illustrated are employed, showing that model building always requires an accurate analysis of the problem by the researcher.

April 10, 2023

Mariangela Guidolin
Padova

Acknowledgments

I could not have undertaken this journey without the ideas, knowledge, and expertise of Renato Guseo. I would like to express my deepest gratitude for his guidance and support in these years of research on innovation diffusion modeling.

I am also grateful to Piero Manfredi for all his valuable and inspiring suggestions and for his constant encouragement during the planning and development of this work.

I am in debt to Andrea Savio, Federico Zanghi, Filippo Ziliotto, and Alessandro Bessi for their enthusiastic work to develop the R package "DIMORA" on which all the analyses contained in this book are based. A special thanks also goes to Carlo De Dominicis for developing the companion website of the book. I would also like to thank all my students in Statistical Sciences and Data Science for giving me constant motivation.

Finally, this book would have been impossible without the guidance, invaluable knowledge, insightful suggestions, constructive criticism, and endless support of Bruno Scarpa.

Acronyms

AIC	Akaike information criterion
ACF	autocorrelation function
ARMA	autoregressive moving-average
ARMAX	autoregressive moving-average with external regressor
BM	Bass model
GBM	generalized Bass model
GBMe1	generalized Bass model with one exponential shock
GBMr1	generalized Bass model with one rectangular shock
GBMe1r1	generalized Bass model with one exponential and one rectangular shock
GGM	Guseo–Guidolin model
GGM-R	Guseo–Guidolin model with one rectangular shock
LVch	Lotka–Volterra with churn model
NLS	nonlinear least squares
UCRCD	unbalanced competition regime change diachronic model
UCTT	unbalanced competition for three technologies
WN	white noise

About the Companion Website

This book is accompanied by a companion website:

www.wiley.com/go/innovationdiffusionmodels/

The website includes:

- Data Sets
- R Codes

Introduction

Given that new products and technologies directly affect many aspects of the lives of people, communities, countries, and economies, the diffusion of innovations appears to be a critical field of research. Modeling and forecasting the diffusion of innovations are broad and relevant fields of study that highlight the importance of innovations in triggering the evolution of social and economic systems and the central role played by diffusion models from a strategic and anticipative point of view. Researchers have observed that "no other field of behavioral science research represents more effort by more scholars in more disciplines in more nations" (Rogers, 2003). The phenomenon of innovation diffusion is essentially a social one, but it has always attracted many researchers from different disciplines, allowing the combination of theories, concepts, and models from many disciplines, including the natural sciences on the one hand, namely mathematics, physics, statistics, and the social sciences on the other, such as marketing science, economics, sociology, and technology management. The mathematical modeling of diffusion processes has been based on epidemic models of biology and mathematical epidemiology, namely, the logistic or s-shaped equation, because it has been observed that innovation spreads in a social system through imitation between people, as a disease spreads in a population through contagion. The most famous and widely-used evolution of this equation is the Bass model (BM) (Bass, 1969), which was developed in the field of quantitative marketing and soon became a major reference because of its surprisingly simple structure on the one hand and its predictive ability on the other. Since the publication of the BM in *Management Science* in 1969, the field of quantitative marketing has led in defining the boundaries and the directions of innovation diffusion theorizing and modeling. One of the characterizing aspects of the BM is that it addresses markets in aggregate; using aggregate adoption data, it depicts and predicts the development of an innovation life cycle already in progress. In strategic terms, crucial forecasts concern the point of maximum growth of the life cycle, the peak, and the point of market saturation. Innovation diffusion models can be used, both in a predictive way and for ex post description, to help understand the evolution of a particular market and its response to various elements, such as marketing strategies, incentive mechanisms, changes in prices, and policies. Timely investigations into the evolving structure of markets for innovations are particularly crucial from an economic and managerial perspective, especially because of the shortening of innovation life cycles, the increasing level of competition between firms and products, and the rise of successive generations of products.

References

Frank M Bass. A new product growth model for consumer durables. *Management Science*, 15(5):215–227, 1969.

Everett M Rogers. *Diffusion of Innovations*. New York: Free Press, 2003.

1

Theory of Innovation Diffusion

After reading this chapter, you should be able to

- Know the basic concepts of innovation and innovation diffusion
- Know what is the purpose of innovation diffusion models
- Know the key principles of innovation diffusion models.

1.1 Basic Concepts and Definitions

1.1.1 Innovation

The term innovation means a new product, a new service, or a new technology. Innovations are new ideas and new ways of doing things. The economist Schumpeter (1947) simply defined innovation as "the doing of new things or the doing of things that are already being done in a new way" and pointed out that innovations are the drivers of economic change. Definitely, innovations in history have directly affected the lives of people, communities, and countries by improving quality of life and bringing progress and wealth. Why are innovations so important? An enormous variety of possible examples in every commercial, industrial, and technological sectors may demonstrate how innovations impact both producers and consumers. On the one hand, innovations can have the possibility to improve the lives of people, allowing technological, socioeconomic, and cultural evolution; on the other hand, they are vital for economies and firms. Firms continuously create new products and services to retain their customers and gain new ones, by generating new market niches. Consumers always have new needs and expectations, deriving from new lifestyles, new cultural trends, and demographic changes. The dynamics of innovation have been at the core

Innovation Diffusion Models: Theory and Practice, First Edition. Mariangela Guidolin.
© 2024 John Wiley & Sons Ltd. Published 2024 by John Wiley & Sons Ltd.
Companion Website: www.wiley.com/go/innovationdiffusionmodels/

of an endless stream of studies, aimed at shedding light on economic, social, technological, demographic, and cultural aspects concerning it (Fagerberg et al., 2005). A typical distinction proposed in innovation studies concerns the level of innovativeness, and innovations are classified as "radical" or "incremental," based on the fact that they are a complete novelty or mere improvements and successive generations of already existing things (Fagerberg et al., 2005). This is a crucial distinction that calls attention on the fact that incremental innovations are easier to realize and adopt, whereas radical innovations are less frequent and create totally new market niches. The sociologist Rogers (2003), in his famous book "The Diffusion of Innovations," suggested that innovations may be described according to five key attributes:

- *Relative advantage*: The level of improvement obtained with the new product with respect to the old one;
- *Compatibility*: The level of coherence of the innovation with existing values, experiences, and needs of consumers;
- *Complexity*: The level of difficulty to understand and use the innovation;
- *Testability*: The possibility to test the innovation for a period of time, reducing the risk connected to it;
- *Observability*: The consequences of the innovation being visible to others.

The analysis of these key factors allows for understanding why some innovations are more difficult to be accepted than others. Indeed, the decision to adopt may require considerable time and cognitive effort: however, most of the time, people do not have sufficient knowledge to evaluate the advantage implied by a new product or technology compared to investment risks, costs, and uncertainty, and acquiring such a piece of knowledge in the short term is difficult. In addition, the complexity of decisions may be due to the fact that the full benefits of adoption are typically delayed with respect to the time of purchase when negative outcomes are more likely to occur. A noteworthy situation is represented by interactive innovations, such as telephones, fax machines, emails, and social networks, which present strong *network externalities* (Katz and Shapiro, 1986). In this case, the benefit of innovation for a single user depends on the number of others who have already adopted it and, thus, on the existence of a physical or virtual network of interacting devices.

All these considerations suggest that the meaning of innovation is fully appreciated if considered within a historical process of acceptance and adoption.

The success of innovation needs to be analyzed with a dynamic, evolutionary perspective that links the innovation to its *diffusion* process. Diffusion is not only the process by which an innovation spreads in a population, it is also an intrinsic part of the innovation process, because learning, imitation, and feedback effects help improve the original innovation (Fagerberg et al., 2005).

1.1.2 Innovation Diffusion

Innovation diffusion has been defined as "the process by which an innovation is communicated through certain channels over time among the members of a social system" (Rogers, 2003). As such, it consists of four central elements:

1. *Innovation*
2. *Communication channels*
3. *Time*
4. *Social system*

As already mentioned, an innovation may be a new product, service, technology, or idea that needs to be accepted from a social system. Communication is the mechanism by which members of a social system create and share information to reach a reciprocal comprehension and a common agreement about the value and the acceptability of the innovation: the structure of communication is essential to understand the speed of a diffusion process. We may distinguish two main forms of communication: institutional communication and interpersonal communication (Rogers, 2003).

Not only does diffusion occur within a given social system, which is a set of interconnected individuals with different roles and relationships, but also through time. Innovation diffusion is an evolutionary process, given that the acceptance of a novelty in a social system is not instantaneous, but sometimes requires a long time to occur. The economist Rosenberg (1976) observed that "in the history of the diffusion of many innovations, one cannot help being struck by two characteristics of the diffusion process: its apparent overall slowness on the one hand, and the wide variations in the rates of acceptance of different inventions, on the other."

1.1.3 Innovation Diffusion Models

The purpose of an innovation diffusion model is to understand this process, which occurs in time and space, and predict its future evolution. Using the definition of Mahajan and Muller (1979), the purpose of an innovation diffusion model is to "depict the successive increases in the number of adopters given a set of potential adopters over time and predict the development of a diffusion process already in progress." Diffusion modeling is important, both for firms that introduce new products and for firms that offer complementary or substitute products. The shortening of product life cycles, the increasing level of competition between firms and products, and the need to plan the likely existence of several successive generations of a new product and thus to manage resources and commitments require a timely investigation of the features of new product growth in terms of its speed and size.

The fundamental marketing concept underlying the employment of innovation diffusion models is that of *product life cycle* (e.g. Day, 1981). The product life cycle hypothesizes that sales of a new product are characterized by stages of launch, growth, maturity, and decline, mimicking the life cycle of a biological organism. The product life cycle may be considered an *empirical generalization* that is "a pattern or regularity that repeats over different circumstances and that can be described simply by mathematical, graphic, or symbolic methods" (Bass, 1995). Each stage of the product life cycle is characterized by well-defined purposes and strategies. In this sense, marketing actions play a special role in trying to stimulate and influence this process. However, the success of innovations depends ultimately on consumers accepting them (Hauser et al., 2006). Innovation diffusion models are especially concerned with this process of adoption.

Figures 1.1 and 1.2 illustrate some examples of innovation diffusion. Figure 1.1 displays annual sales of compact discs in the United States (in million units) from 1982 to 2021, whereas Figure 1.2 shows the quarterly sales of Apple iPod (in millions, at the world level) from the first quarter of 2002, Q1'02, to the fourth quarter of 2014, Q4'14. Apart from evident differences due to the presence of a strong seasonality in the iPod case, these two technologies share a common nonlinear

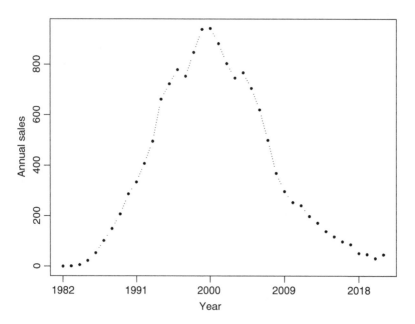

Figure 1.1 Annual sales of compact discs in the United States.

path, characterized by the phases of introduction, growth, maturity, and decline, coherent with the life cycle concept. Clearly, these two examples have especially an illustrative purpose because the life cycle is essentially complete; therefore, the application of a model has more descriptive power rather than predictive.

A different, and perhaps more challenging, case is that represented in Figure 1.3, where the diffusion process of wind energy in Germany is displayed, from 1993 to 2020. The adoption of new technology, such as renewable energy technology, may be taken as a paradigmatic case of a crucial innovation diffusion process, characterized by barriers and difficulties, but also with a great potential of changing and improving the lives of people, as described by Rogers (2003). In this case, the ability to predict the future evolution of the diffusion process is more critical because of the uncertainty related to the process on the one hand and the importance of reliable predictive models on the other. As may be seen in Figure 1.3, the process has a clearly growing trend and the life cycle of the technology could be very long.

This book is an introduction to a class innovation diffusion models that can be used to describe and forecast the evolution in time of sales of new products or technologies. Starting from the basic Bass model (BM), the book presents some of its generalizations, which account for the presence of exogenous shocks, affecting the timing of the diffusion process, and for the presence of a dynamic market potential, as a function of a communication process, which develops over time. Moreover, some generalizations of the univariate BM are proposed to account for the presence of competition.

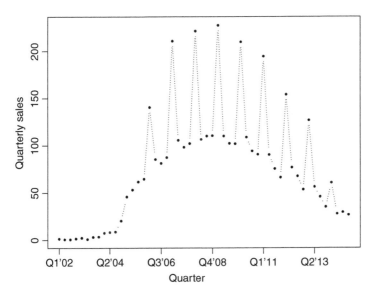

Figure 1.2 Quarterly sales of Apple iPod.

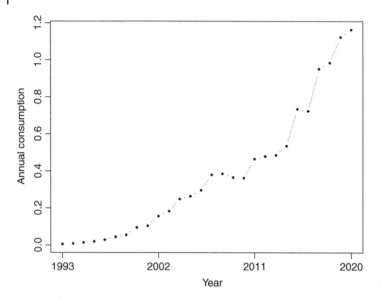

Figure 1.3 Annual consumption of wind energy in Germany.

The statistical techniques involved in model estimation combine time-series analysis with nonlinear regression techniques.

The key objectives of the book are:

- to describe the main mathematical features of the models, discussing the meaning of the parameters from the marketing point of view with several real-data examples,
- to present and discuss the statistical aspects involved in model estimation and selection,
- to show and discuss the predictive and explanatory ability of the proposed models with real-data applications in commercial and economic applications, highlighting the properties and limitations of each of the models described,
- to propose some modeling advancements for future achievements in research and practice.

The book does not intend to cover the wide variety of models proposed in the extremely rich literature on innovation diffusion modeling, which has expanded in several directions over the years. Reference books and reviews of this literature may be found, for example, in Mahajan et al. (1990, 2000), Meade and Islam (2006), Chandrasekaran and Tellis (2007), Muller et al. (2009), Peres et al. (2010), and Guidolin and Manfredi (2023). For a general description of growth models from a statistical point of view, a reference book is Seber and Wild (1989).

Rather, this book takes a specific perspective aimed at discussing and proposing a solution to some of the well-known limitations of the BM, to provide a possibly better description of market penetration dynamics. For each of the generalizations discussed, some real-data applications are presented, to show the usefulness and usability of the models. In so doing, some key principles have been especially accounted for, namely,

1. *Parsimony*: The model should be statistically identifiable and significant,
2. *Flexibility*: The model should be able to account for different diffusion patterns,
3. *Practical usability*: The model should be easy to use, with the only availability of aggregate data,
4. *Interpretability*: The model should be easy to interpret and all the parameters should have a clear meaning in commercial or technological terms.

The book has been developed from a general perspective, to be employed with various possible statistical software. However, all the analyses, outputs, and graphs here contained have been produced in the R environment (R Core Team, 2021), by using the package DIMORA (Zanghi et al., 2021) for innovation diffusion models.

References

Frank M Bass. Empirical generalizations and marketing science: A personal view. *Marketing Science*, 14(3 supplement):G6–G19, 1995.

Deepa Chandrasekaran and Gerard J Tellis. A critical review of marketing research on diffusion of new products. *Review of Marketing Research*, 3:39–80, 2007.

George S Day. The product life cycle: Analysis and applications issues. *Journal of Marketing*, 45(4):60–67, 1981.

Jan Fagerberg, David C Mowery, Richard R Nelson, et al. *The Oxford Handbook of Innovation*. Oxford: Oxford University Press, 2005.

Mariangela Guidolin and Piero Manfredi. Innovation diffusion processes: Concepts, models, and predictions. *Annual Review of Statistics and Its Application*, 10(1):451–473, 2023. doi: 10.1146/annurev-statistics-040220-091526.

John Hauser, Gerard J Tellis, and Abbie Griffin. Research on innovation: A review and agenda for marketing science. *Marketing Science*, 25(6):687–717, 2006.

Michael L Katz and Carl Shapiro. Technology adoption in the presence of network externalities. *Journal of Political Economy*, 94(4):822–841, 1986.

Vijay Mahajan and Eitan Muller. Innovation diffusion and new product growth models in marketing. *Journal of Marketing*, 43(4):55–68, 1979.

Vijay Mahajan, Eitan Muller, and Frank M Bass. New product diffusion models in marketing: A review and directions for research. *Journal of Marketing*, 54(1):1–26, 1990.

Vijay Mahajan, Eitan Muller, and Yoram Wind. *New-Product Diffusion Models*, volume 11. New York: Springer Science & Business Media, 2000.

Nigel Meade and Towhidul Islam. Modelling and forecasting the diffusion of innovation–a 25-year review. *International Journal of Forecasting*, 22(3):519–545, 2006.

Eitan Muller, Renana Peres, and Vijay Mahajan. *Innovation Diffusion and New Product Growth*. Cambridge: Marketing Science Institute, 2009.

Renana Peres, Eitan Muller, and Vijay Mahajan. Innovation diffusion and new product growth models: A critical review and research directions. *International Journal of Research in Marketing*, 27(2):91–106, 2010.

R Core Team. *R: A Language and Environment for Statistical Computing*. R Foundation for Statistical Computing, Vienna, Austria, 2021. URL https://www.R-project.org/.

Everett M Rogers. *Diffusion of Innovations*. New York: Free Press, 2003.

Nathan Rosenberg. *Perspectives on Technology*. Cambridge: Cambridge University Press, 1976.

Joseph A Schumpeter. The creative response in economic history. *The Journal of Economic History*, 7(2):149–159, 1947.

George AF Seber and Chris J Wild. *Nonlinear Regression*. New York: Wiley, 1989.

Federico Zanghi, Andrea Savio, Filippo Ziliotto, et al. *DIMORA: Diffusion Models R Analysis*, 2021. URL https://CRAN.R-project.org/package=DIMORA. R package version 0.2.0.

2

Innovation Diffusion as an Empirical Generalization: The Bass Model

After reading this chapter, you should be able to

- Know the basic features of the Bass Model
- Interpret the parameters of the Bass Model
- Make predictions with the Bass Model.

2.1 Introduction

One of the key factors for innovation diffusion processes is the existence and strength of some communication channels, that is, the means by which information about innovation is transmitted to or within the social system (Mahajan et al., 1990). These communication channels may be both formal, such as mass media and advertising, and informal, such as interpersonal communication and learning from others' experiences. In particular, "interpersonal communications, including nonverbal observations, are important influences in determining the speed and shape of the diffusion process in a social system" (Mahajan et al., 1990).

The mathematical description of innovation diffusion has typically used epidemic models borrowed from biology, namely, the logistic equation, in which the social contagion represents the driving factor of growth. The logistic equation was developed for the first time by Verhulst (1838) and was originally used in natural sciences for describing growth processes, such as the spread of a disease or the growth of a biological organism. The application of this model in the context of new products, technologies, and services relies on the hypothesis that innovation spreads in a market through interpersonal communication and imitation like an epidemic disease through the mechanism of contagion among individuals.

Innovation Diffusion Models: Theory and Practice, First Edition. Mariangela Guidolin.
© 2024 John Wiley & Sons Ltd. Published 2024 by John Wiley & Sons Ltd.
Companion Website: www.wiley.com/go/innovationdiffusionmodels/

The central role of imitation in explaining diffusion processes and the possibility to represent them with the logistic curve are common elements of many innovation diffusion models and applications.

The logistic equation is defined with the first-order differential equation

$$f'(t) = k\frac{f(t)}{\alpha}[\alpha - f(t)], \quad t > 0,$$

where $f'(t)$ is the first derivative of $f(t)$. The growth rate $f'(t)$ is proportional to the residual *carrying capacity*, $\alpha - f(t)$, multiplied by the term $kf(t)/\alpha$, which expresses the contagion factor.

The marketing field has become central in innovation diffusion modeling since the 1970s. Pioneering contributions in this area are those of Mansfield (1961), reproducing the Verhulst structure, and Bass (1969). The Bass model (BM) has inspired an extremely productive stream of research, dealing with the modeling and forecasting of innovation diffusion. The BM has been primarily developed for studying the life cycle of durable goods, but its application has later been found to be suitable for many other commercial and technological sectors, such as non-durable goods, services, and technologies.

In this chapter, specific attention is devoted to the theoretical description and practical application of the BM. Section 2.2 describes the basic mathematical features of the model, illustrating the meaning of the parameters and how the diffusion pattern depends on these. Section 2.3 illustrates some key aspects referred to as model estimation, the goodness of fit and model selection. Section 2.4 discusses the application of the model to some real datasets. The strengths and weaknesses of the BM are summarized in Section 2.5.

2.2 Bass Model: Theory

The BM describes the life cycle of an innovation, depicting its characteristic phases of launch, growth and maturity, and decline. As mentioned before, the model was originally developed in marketing science and its purpose is to model the growth of a new product over time as a result of the purchases of a given set of potential adopters. The model assumes that these purchases are influenced by two sources of information, an external one, such as mass media and advertising, and an internal one, namely, imitation and learning from others. These different sources of information create two groups of adopters. One group is influenced only by the external source, the *innovators*, and the other only by the internal one, and these are the *imitators*. One of the great advantages associated with the BM is the concrete possibility to explain the initializing phase of diffusion due to the presence of innovators. Indeed, a huge body of literature is available regarding the role of innovators, also called *early adopters* by Rogers (2003). However, the first model explicitly accounting for their role is the BM. In particular, a constant level of innovators buying the product at the beginning of the diffusion is assumed to exist. In this sense, the BM accounts for the role of all the communication efforts realized by firms, whereas a pure logistic approach like that of the Mansfield model does not.

The formal representation of the BM is a first-order differential equation

$$z'(t) = \left[p + q\frac{z(t)}{m} \right] [m - z(t)], \quad t > 0. \tag{2.1}$$

In Equation (2.1), the variation over time of adoptions, $z'(t)$, is proportional to the residual market, $m - z(t)$, where m is the market potential, and $z(t)$ is the cumulative number of adoptions at time t. The market potential m is the maximum number of realizable sales within the diffusion, and its value is assumed to be *constant* throughout the entire process. The residual market, $m - z(t)$, is affected by the coefficients p and q. Parameter p, called *innovation coefficient*, represents the effect of the external influence, due to the mass media communication and advertising, whereas parameter q, called *imitation coefficient*, is the internal influence, whose effect is modulated by the ratio $z(t)/m$.

Parameters p and q measure the two different categories of consumers mentioned before, the innovators and the imitators. The innovators are those that adopt the innovation first and on the basis of a personal belief developed through media communication, whereas the imitators are those adopting at a second stage, by imitating the behavior of others, as expressed through the component qz/m, termed *word-of-mouth*.

Note that $qz/m = 0$ as long as cumulative adoptions are zero, i.e. $z(t) = 0$, and this justifies the fact that imitators start to buy at a later stage, given that cumulative sales need to be positive, i.e. $z(t) > 0$. When the product enters the market, at time $t = 0$, there are no cumulative sales, $z(t) = 0$. Therefore Equation (2.1) is reduced to $z'(0) = pm$. Hence, at the beginning of diffusion, there is a constant level of innovators pm, which represents the positive starting point of the process.

By posing $z(t)/m = y(t)$, the model may be re-written as

$$y'(t) = [p + qy(t)][1 - y(t)], \quad t > 0, \tag{2.2}$$

where y may be seen as a cumulative distribution function, and y' is the corresponding density function. Equation (2.2) may be rearranged to express the BM as a hazard function, that is, the probability that an event will occur at time t, given that it has not occurred yet

$$\frac{y'}{(1 - y)} = p + qy, \quad t > 0. \tag{2.3}$$

Equation (2.3) describes the conditional probability of adoption at time t, resulting from the sum of the probabilities of two incompatible events, p and qz/m.

Table 2.1 Bass model: parameters and description.

Parameter	Description
m	Market potential
p	Innovation
q	Imitation

This points out that the model does not consider adoptions due to both innovative and imitative behavior (Guseo, 2004). Table 2.1 summarizes the parameters of the BM with their meaning.

2.2.1 Closed-Form Solution

The differential Equation (2.2) has a closed-form solution (Bass, 1969):

$$y(t) = \frac{1 - e^{-(p+q)t}}{1 + \frac{q}{p}e^{-(p+q)t}}, \quad t > 0. \tag{2.4}$$

In Equation (2.4), function $y(t)$ may take values between 0 and 1, $0 < y(t) < 1$, and directly depends on parameters p and q, whose magnitude determines the speed of growth, until saturation.

Furthermore, because $z(t) = y(t)m$, the closed-form solution of the BM may be rewritten as

$$z(t) = m\,\frac{1 - e^{-(p+q)t}}{1 + \frac{q}{p}e^{-(p+q)t}} = my(t), \quad t > 0. \tag{2.5}$$

In Equation (2.5), the dynamics of the diffusion process in terms of cumulative sales, $z(t)$, are described with three parameters: m, p, and q. As in Equation (2.4), parameters p and q act on the speed of diffusion, whereas the market potential m is a scale parameter and allows for modeling the diffusion process in absolute terms.

Figure 2.1 describes the behavior of cumulative adoptions, $z(t)$, as expressed in Equation (2.5), for some values of parameters p, and q. As one may easily observe, the cumulative process has an *s-shaped* pattern and reaches the saturation level, represented by the parameter m, with different speeds depending on the value taken by parameters p and q. Higher values of p and q, imply a faster diffusion process: for example, if $p = 0.01$ and $q = 0.1$, the diffusion process will reach the saturation level m more rapidly.

The corresponding instantaneous adoptions, $z'(t)$, are defined as follows

$$z'(t) = m\,\frac{p(p+q)^2 e^{-(p+q)t}}{[p + qe^{-(p+q)t}]^2}, \quad t > 0. \tag{2.6}$$

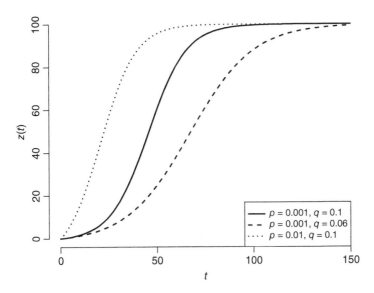

Figure 2.1 Three cumulative diffusion processes, $z(t)$, with different p and q parameters. The market potential m is fixed at $m = 100$.

Instantaneous adoptions, $z'(t)$, of Equation (2.6) are displayed in Figure 2.2, and highlight the presence of a *maximum peak*, which is the maximum instantaneous expansion of the diffusion process. In particular, the maximum peak is reached at time t^*

$$t^* = \frac{\log(q/p)}{(p+q)}, \tag{2.7}$$

where function $z(t)$ takes the value

$$z(t^*) = \frac{m}{2} - \frac{p}{2q}. \tag{2.8}$$

From Equation (2.8), given that $p/2q$ typically takes a negligible value, one may conclude that $z(t^*) \approx m/2$, that is, when half of the market potential has been reached. In Figure 2.2 one may easily notice that with higher values of p and q, the maximum peak is reached more rapidly, while lower parameters of p and q will imply a slower diffusion process.

In strategic terms, the peak represents a crucial moment, given that the diffusion process starts declining after this. From a product life cycle and marketing perspective, the decline phase after the peak needs to be carefully managed by the firm, for example, by developing successive generations of the product.

Figure 2.2 also highlights that the diffusion processes start from a positive level: this is the initialization level represented by pm, which characterizes the structure of the BM.

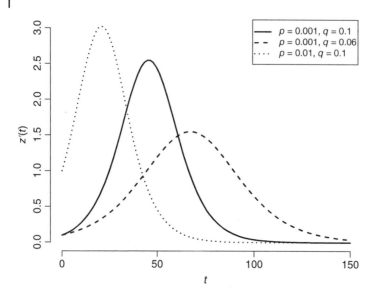

Figure 2.2 Three instantaneous diffusion processes with different p and q parameters. The market potential m is fixed at $m = 100$.

2.3 Model Estimation

A nonlinear least squares (NLS) method may be used to fit a BM to the available data (Guseo, 2004). Following Seber and Wild (1989), we may consider the structure of a nonlinear model, as follows:

$$w(t) = \eta(\beta, t) + \varepsilon(t),$$

where $w(t)$ is the observed response; $\eta(\beta, t)$ is the deterministic component describing the instantaneous or cumulative process, $z'(t)$ or $z(t)$, depending on parameter set β and time t, and $\varepsilon(t)$ is an error term that is assumed to be a *white noise* process. In the case of the BM, $\eta(\beta, t)$ will be represented by Equation (2.5) or (2.6). In general, as is the case for the BM and most models we will see in chapters 3 and 4, whenever a closed-form solution is available, this will be employed for model fit on cumulative data $z(t)$.

The solution of NLS is based on numerical algorithms, such as Gauss–Newton or Levenberg–Marquardt (for the details, see Chapter 7). The use of a numerical algorithm requires the preliminary specification of some starting values for parameters: for BM we need starting values for m, p, and q. As we have seen in describing the BM, the number of innovators is typically much lower than that of imitators, and to reflect this aspect typical starting values for p and q are 0.001 and 0.1.

Moreover, parameter estimation is very sensitive to the number of available data. In particular, estimating m is especially difficult when little data are available. In fact, reliable estimates are obtained after the maximum peak, although it is quite obvious that "by the time sufficient observations have been developed for reliable estimation, it is too late to use the estimates for forecasting purposes" (Mahajan et al., 1990). Thus, there is a clear trade-off between the reliability of estimates and the usefulness of the model for prediction.

2.3.1 Goodness of Fit

Model global *goodness of fit* is evaluated through the R^2,

$$R^2 = 1 - \frac{\text{residual deviance}}{\text{total deviance}}$$

the value of which is typically greater than 0.95, when the fit is calculated on cumulative data so that it is often not a sufficient indicator of model performance. From this perspective, comparison and selection between different but nested models become especially important. In regression settings, the performance of an extended model, m_2, compared with a nested one, m_1, may be evaluated through a squared multiple partial correlation coefficient $\tilde{R}^2$ in the interval $[0; 1]$, namely,

$$\tilde{R}^2 = (R^2_{m_2} - R^2_{m_1})/(1 - R^2_{m_1}),$$

where $R^2_{m_i}$, $i = 1, 2$, is the determination coefficient of model m_i.

The $\tilde{R}^2$ coefficient has a monotone correspondence with the F ratio

$$F = \frac{\tilde{R}^2(n - v)}{(1 - \tilde{R}^2)u},$$

where n is the number of observations, v is the number of parameters of the extended model m_2, and u is the incremental number of parameters from m_1 to m_2. Conditional to the distribution of the error term $\varepsilon(t)$, the statistic F, for the null hypothesis of equivalence of the two models, is a central Snedecor's F with u degrees of freedom for the numerator and $n - v$ degrees of freedom for the denominator, $F \sim F_{u,n-v}$ (Guseo et al., 2007, Guidolin and Mortarino, 2010). As a rule of thumb, an extended model, m_2, may be considered suitable whenever $\tilde{R}^2 \geqslant 0.3$ and the $F \geqslant 4$ (Guseo, 2004).

2.4 The Bass Model: Case Studies

This section is dedicated to the application of the BM to sales data at the world level of two products pertaining to the ICT sector, the Apple iPhone and the RIM

Blackberry. For both applications, the results of the BM are discussed both from a statistical and marketing point of view. The data employed have been taken from www.apple.com and www.statista.com (for Blackberry).

A third application is referred to the adoption of wind energy in Denmark, taken as a paradigmatic example of similar diffusion processes occurring in other countries. The data are based on the BP Statistical Review of World Energy 2021 available at www.bp.com.

2.4.1 Model Fit

The estimation procedure, based on NLS through the Levenberg–Marquardt algorithm (see Chapter 7), requires specifying some starting values for parameters m, p, and q. A rule of thumb suggests specifying for m a slightly higher value than that observed for cumulative data $z(t)$, while parameters p and q have been, respectively, set $p = 0.001$ and $q = 0.1$.

The applied model is therefore

$$w(t) = m \, \frac{1 - e^{-(p+q)t}}{1 + \frac{q}{p} e^{-(p+q)t}} + \varepsilon(t),$$

where $w(t)$ are observed cumulative data, m, p, and q are the parameters to be estimated, and $\varepsilon(t)$ is the error term assumed to be a white noise process. However, such assumption in many applied cases is not verified and suitable management of regression residuals is required (for details see Chapter 5).

2.4.2 Apple iPhone

Figure 2.3 displays the time series of quarterly sales of the iPhone from its launch in the market in the third quarter of 2007, Q3'07, to the end of the year 2018 (after this year Apple has stopped publicly reporting sales data). Before entering in more detail, it is important to observe that the data considered refer to all generations of iPhone, with no distinction, so that they include repeated purchases. The first inspection of data in Figure 2.3a suggests some preliminary considerations: there is evidence of a nonlinear trend of the series, with a clear growth until the year 2016 and a subsequent slowdown in the last two years, and of a very strong seasonal pattern with a peak every first quarter, especially visible from 2012. The BM is not developed to handle such seasonal behavior and this will be ignored for the moment while focusing just on the nonlinear trend. A specific treatment of seasonality is proposed in Chapter 5.

As already observed, according to the established literature on diffusion modeling, the most accepted estimation procedure of the BM is performed by taking advantage of the closed-form solution expressed in Equation (2.5), requiring the

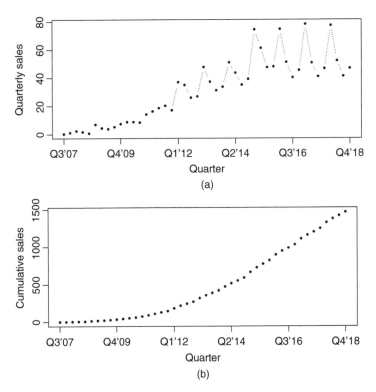

Figure 2.3 iPhone. Quarterly (a) and cumulative (b) sales from 2007 to 2018 (million units).

use of cumulative data. The first step is therefore to consider cumulative data, displayed in Figure 2.3b. Cumulative data well show an s-shaped pattern consistent with that illustrated in Figure 2.1.

The results of model fitting are outlined in Table 2.2, where estimates, standard errors, and 95% confidence intervals are reported for m, p, and q. As a general remark, if confidence intervals do not contain the zero value, this indicates that the parameter is statistically significant at 5%.

An indication of goodness of fit is provided through the $R^2 = 0.9995$. As mentioned earlier, the value of the R^2 in these applications usually takes very high levels, since it is based on cumulative data. Figure 2.4 shows the fit of the BM to cumulative data and the corresponding fit to instantaneous (quarterly) data. According to the prediction performed with the BM, the maximum peak in sales was reached in 2016, followed by a declining phase, which would predict the market exit in a few years. Although the BM may be employed for making predictions about the future evolution of a product's life cycle, one of its merits is the

Table 2.2   Parameter estimates of BM for iPhone.

Parameter	Estimate	s.e.	Lower c.i.	Upper c.i.
m	1823.7	34.1	1757.9	1890.6
p	0.0014	0.00005	0.0013	0.0015
q	0.13	0.003	0.12	0.13

$R^2 = 0.9995$.

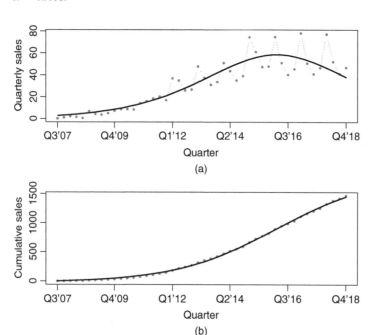

Figure 2.4   BM for iPhone. Quarterly sales (a) and cumulative sales (b).

possibility to have a clear interpretation of the involved parameters from a marketing perspective. The estimated market potential, $m = 1823.7$, represents the maximum number of realizable sales over the entire life cycle of the iPhone. Parameters p and q refer to the innovative and imitative behavior within the diffusion process. It may be observed that parameter $p = 0.0014$ is a hundred times smaller than parameter $q = 0.13$: this is a typical feature of many diffusion processes, where the majority of purchases are made by consumers imitating the choice of others, through the mechanism of word-of-mouth. In this sense, the interpretation of parameters p and q may provide useful insights into the characteristics of a product and its market.

2.4.3 RIM Blackberry

The second product considered is the Blackberry for which quarterly sales data are available from 2007 to 2013. A preliminary analysis of these data, presented in Figure 2.5 shows that we are dealing with a complete life cycle. Obviously, in this case, the application of the BM has just a descriptive purpose.

The estimation strategy follows the phases already described in Section 2.4.2. Results of the application of the BM are presented in Figure 2.6 and Table 2.3. From a general perspective, the BM appears very suitable to describe the nonlinear trend of the series and, despite some irregularities likely due to a seasonal pattern, the maximum peak is efficiently captured. The only data points where the BM shows a worse performance are the last three referred to the year 2013: indeed, the predicted series is not able to describe the accelerated decline in the observed data. This rapid closure of the life cycle may be due to market factors such as the effect of a strong competition from alternative technologies, like the iPhone and other smartphones.

An interesting comparison with the iPhone case may be performed with reference to the estimated parameters. The market potential, $m = 242.4$, is significantly smaller than that of the iPhone, suggesting that Blackberry is indeed a product for

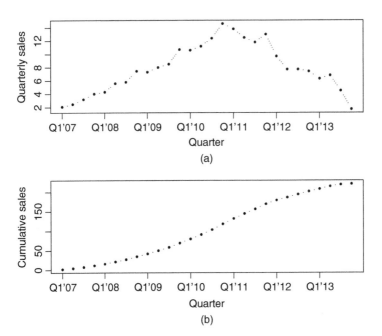

Figure 2.5 Blackberry. Quarterly (a) and cumulative (b) sales from 2007 to 2013 (million units).

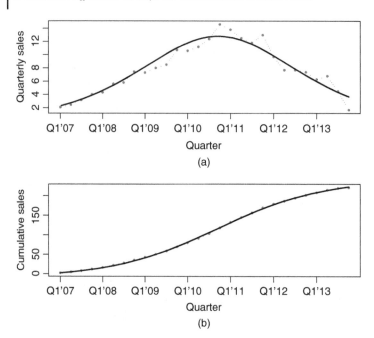

Figure 2.6 BM for Blackberry. Quarterly sales (a) and cumulative sales (b).

Table 2.3 Parameter estimates of BM for Blackberry.

Parameter	Estimate	s.e.	Lower c.i.	Upper c.i.
m	242.4	1.47	239.5	245.3
p	0.0079	0.0002	0.0075	0.0082
q	0.19	0.003	0.19	0.20

$R^2 = 0.9999$.

a more limited market niche. As concerns diffusion parameters, p and q, it may be observed that the life cycle of Blackberry is characterized by a higher innovation coefficient, $p = 0.0079$, which indicates a stronger role for innovators. This finding appears consistent with the hypothesis of a niche product.

2.4.4 Wind Energy Consumption in Denmark

A third example of the application of the BM concerns the diffusion of wind energy. Renewable energies are at the center of a technological revolution, given that their diffusion is a necessary step for the so-called "energy transition." Starting from the pioneering work of Marchetti (1980), the diffusion of renewable energy sources has been the object in recent years of a growing stream of research that employed innovation diffusion models to describe and predict these evolutionary processes (for a review of this literature see, for instance, Petropoulos et al., 2022, p. 786). One may wonder whether renewable energy technology may be treated with a model for the product life cycle. Based on the historical pattern of past energy technologies, as shown for example in Marchetti (1980), we may hypothesize that energy sources are similar to commercial products that need to be accepted in a market niche and are characterized by a finite life cycle, because of competition and substitution dynamics. As in other markets, also in the energy market, new technologies for production and consumption tend to substitute older technologies that progressively decline and finally exit the market.

In this example, the case of Denmark is specifically considered, because of its leading role in the development and usage of wind energy. Figure 2.7 illustrates the annual wind energy consumption from 1990 to 2020 and the corresponding cumulative series.

The trajectory described in Figure 2.7b confirms a behavior for which the BM appears to be a suitable solution. However, we shall observe that the situation presented in this third application is somehow different from those referred to

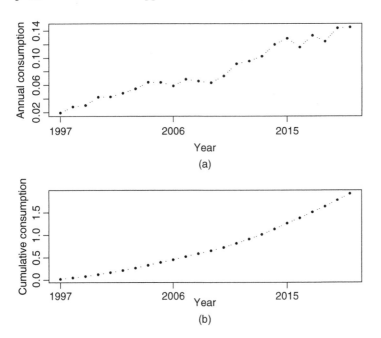

Figure 2.7 Consumption of wind energy in Denmark. Annual (a) and cumulative (b) data from 1997 to 2020 (in ExaJoule).

the iPhone and Blackberry, in terms of the stage of the diffusion process. Whereas in previous examples the data had substantially overcome the maximum peak and, in the case of Blackberry, the life cycle was essentially closed, wind energy is clearly in its growth phase, and no evidence of a maximum peak in the data has been observed. This makes the application of the BM potentially more useful for predictive purposes but may also pose some challenges in terms of parameter estimation. The results of the estimation of the BM are displayed in Figure 2.8 and Table 2.4. Overall, the BM obtains a very satisfactory performance with $R^2 = 0.9998$, and all the parameters are statistically significant, as the inspection

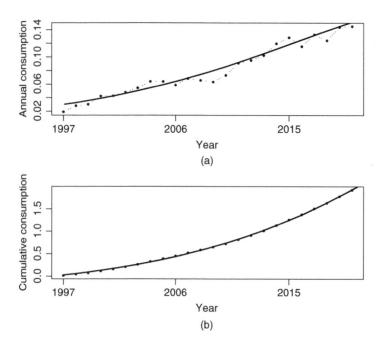

Figure 2.8 BM for wind energy consumption in Denmark. Annual (a) and cumulative (b).

Table 2.4 Parameter estimates of BM for consumption of wind energy in Denmark.

Parameter	Estimate	s.e.	Lower c.i.	Upper c.i.
m	7.24	1.38	4.54	9.94
p	0.004	0.0006	0.002	0.005
q	0.09	0.005	0.08	0.10

$R^2 = 0.9998$.

of Table 2.4 suggests. The role of innovators appears somehow fragile, $p = 0.004$, a phenomenon already documented in the literature on the diffusion of renewable energy sources (Guidolin and Mortarino, 2010, Bunea et al., 2020, 2022). This may justify the difficulties related to the adoption of new energy technology. The graphical visualization of the results suggests that the BM is a suitable model for this series, although some residual variability around the nonlinear mean trajectory may still be observed.

2.5 Recap

At the end of the chapter, some important properties and limitations of the BM are summarized. The use of the BM underlies a strong hypothesis on a product's evolution because it is suitable for products and technologies for which it is reasonable to assume a limited life cycle characterized by phases of birth, growth, maturity, and decline.

Properties

- The BM is a simple yet powerful model, made of only three parameters, m, p, and q, which can effectively describe the diffusion process of a new product, predicting the various phases of its life cycle, that is, growth, maturity (maximum peak), and decline.
- The estimation procedure is based on aggregate time series of sales or adoptions so that in terms of data requirements the BM can be easily employed.
- Parameters have an easy and clear interpretation from a marketing point of view and may lead to a better understanding of a product's market.

Limitations

- The BM can be estimated only when sufficient data are available. Too short time series imply unreliable estimates, whereas the lack of data makes the use of the BM unfeasible. Indeed, this model cannot be used for prelaunch prediction, although some simulations on the potential behavior of a new product could be performed with the BM, by assuming that a new product will behave similarly to other existing products, for which parameters p and q have been already estimated.
- The market potential m is assumed to be a constant over the entire life cycle. In particular, assuming a constant market potential means that a set of potential consumers are ready to buy the product as soon as it is launched into the market. This hypothesis may be reasonable in some situations but appears too restrictive

in others, where it is more reasonable to think of a market potential that needs time to be developed.

- The BM does not account for marketing mix strategies, such as pricing or advertising. Indeed, according to the BM, the only forces driving the diffusion process appear to be the purchase decisions of consumers, both innovators, and imitators. This may suggest two problematic points of the BM: First, the BM appears to neglect the role of marketing in stimulating market growth; second, the BM describes diffusion as a smooth process, whereas real data are generally much more complex and exhibit patterns that cannot be easily captured with a simple BM. A possible solution to overcome such limitation is represented by the Generalized Bass Model, by Bass et al. (1994).

References

Frank M Bass. A new product growth model for consumer durables. *Management Science*, 15(5):215–227, 1969.

Frank M Bass, Trichy V Krishnan, and Dipak C Jain. Why the Bass model fits without decision variables. *Marketing Science*, 13(3):203–223, 1994.

Anita M Bunea, Pompeo Della Posta, Mariangela Guidolin, and Piero Manfredi. What do adoption patterns of solar panels observed so far tell about governments' incentive? Insights from diffusion models. *Technological Forecasting and Social Change*, 160:120240, 2020.

Anita M Bunea, Mariangela Guidolin, Piero Manfredi, and Pompeo Della Posta. Diffusion of solar PV energy in the UK: A comparison of sectoral patterns. *Forecasting*, 4(2):456–476, 2022.

Mariangela Guidolin and Cinzia Mortarino. Cross-country diffusion of photovoltaic systems: Modelling choices and forecasts for national adoption patterns. *Technological Forecasting and Social Change*, 77(2):279–296, 2010.

Renato Guseo. Strategic interventions and competitive aspects in innovation life cycle. Technical report, Department of Statistical Sciences, University of Padua, 2004.

Renato Guseo, Alessandra Dalla Valle, and Mariangela Guidolin. World Oil Depletion Models: Price effects compared with strategic or technological interventions. *Technological Forecasting and Social Change*, 74(4):452–469, 2007.

Vijay Mahajan, Eitan Muller, and Frank M Bass. New product diffusion models in marketing: A review and directions for research. *Journal of Marketing*, 54(1):1–26, 1990.

Edwin Mansfield. Technical change and the rate of imitation. *Econometrica: Journal of the Econometric Society*, 29(4):741–766, 1961.

Cesare Marchetti. Society as a learning system: Discovery, invention, and innovation cycles revisited. *Technological Forecasting and Social Change*, 18(4):267–282, 1980.

Fotios Petropoulos, Daniele Apiletti, Vassilios Assimakopoulos, et al. Forecasting: Theory and practice. *International Journal of Forecasting*, 38(3):705–871, 2022.

Everett M Rogers. *Diffusion of Innovations*. New York: Free Press, 2003.

George AF Seber and Chris J Wild. *Nonlinear Regression*. New York: Wiley, 1989.

Pierre-François Verhulst. Notice sur la loi que la population suit dans son accroissement. *Correspondence Mathematique et Physique*, 10:113–126, 1838.

3

Innovation Diffusion with Structured Shocks: The Generalized Bass Model

After reading this chapter, you should be able to

- Know the basic features of the Generalized Bass Model
- Interpret the parameters of the Generalized Bass Model
- Make predictions with the Generalized Bass Model.

3.1 Introduction

Reviews on diffusion models, such as that proposed by Mahajan et al. (1990), pointed out that a significant limitation of the Bass model (BM) was the failure to incorporate marketing mix variables in the model under managerial control, such as price strategies and advertising. Peres et al. (2010) observed that this omission has raised a conceptual conflict. This is because the model provides a high level of fit and reliable forecasts just making some hypotheses about the behavior of consumers and without marketing mix variables, but at the same time, it is clear that marketing mix decisions exert a notable impact on new product growth. In addition, the shortening of life cycles due to successive generations, analyzed in Norton and Bass (1987), and faster consumption dynamics due to the power of word-of-mouth, have increased the need for a model incorporating control variables. Bass (1995) proposed a synthesis of several attempts to introduce control variables into diffusion models and listed some desirable properties of a diffusion model with decision variables.

This model should:

1. Have empirical support and be managerially useful, allowing a direct interpretation of parameters and comparisons with other situations;

Innovation Diffusion Models: Theory and Practice, First Edition. Mariangela Guidolin.
© 2024 John Wiley & Sons Ltd. Published 2024 by John Wiley & Sons Ltd.
Companion Website: www.wiley.com/go/innovationdiffusionmodels/

2. Have a closed-form solution;
3. Be easy to implement.

A model satisfying all these requirements, formalized by Bass et al. (1994), is the Generalized Bass model (GBM).

The rest of the chapter is structured as follows. Section 3.2 describes the basic structure of the GBM, with some specifications of structured shocks, which are useful in applied cases. Section 3.3 discusses some applications to real-world data. Section 3.4 summarizes the properties and limitations of the GBM.

3.2 Generalized Bass Model: Theory

Conceived for taking into account both price and advertising strategies, the GBM enlarges the basic structure of the BM by multiplying its basic structure by a very general intervention function $x(t)$, assumed to be nonnegative and integrable. The GBM has the following simple structure

$$z'(t) = \left[p + q\frac{z(t)}{m} \right] [m - z(t)]x(t), \quad t > 0, \quad x(t) > 0. \tag{3.1}$$

In Equation (3.1), the original form of function $x(t)$ was designed by Bass et al. (1994), jointly considering the percentage variation of prices and advertising efforts, had the form $x(t) = 1 + \beta_1 Pr'(t)/Pr(t) + \beta_2 A'(t)/A(t)$, where $Pr(t)$ and $A(t)$ are price and advertising at time t, and $Pr'(t)$ and $A'(t)$ are the rates of change in price and advertising, respectively, at time t.

One interesting feature of the GBM is that it reduces to the BM, when $x(t) = 1$, that is when no changes occur in price and advertising. If the percentage changes in price and advertising remain the same from one period to the next, then function $x(t)$ reduces to a constant, again yielding the BM. This would provide a reason why the BM provides good parameter estimates, even without accounting for marketing mix variables. This generalization allows testing the effect of marketing mix strategies on diffusion and making scenario simulations based on the modulation of the function $x(t)$.

As a general note, if $0 < x(t) < 1$, the diffusion process slows down, whereas if $x(t) > 1$, the diffusion accelerates.

3.2.1 Closed-Form Solution

One of the important properties of the GBM is to have a closed-form solution (Bass et al., 1994)

$$z(t) = m \frac{1 - e^{-(p+q)\int_0^t x(\tau)d\tau}}{1 + \frac{q}{p}e^{-(p+q)\int_0^t x(\tau)d\tau}}, \quad t > 0. \tag{3.2}$$

As Equation (3.2) shows, the model's internal parameters, m, p, and q, are not modified by function $x(t)$ (Bass et al., 1994). In fact, function $x(t)$ acts on the shape of diffusion, modifying its development over time and not the value of the parameters, m, p, and q: hence, function $x(t)$ may represent not only price and advertising, but all the strategies that are set to control the timing of a diffusion process, by anticipating or delaying adoptions. Before entering into the details on how to specify $x(t)$, we remind that the GBM reduces to the BM when $x(t) = 1$. This means that the BM is a special case of GBM, allowing a direct comparison of the two models, in terms of performance, specifically through the $\tilde{R}^2$ index and the corresponding F test. The $\tilde{R}^2$ index allows evaluating whether the application of the GBM implies a significant improvement in terms of reduction of residual sum of squares with respect to the simple BM. This also provides an indication of the importance of including an $x(t)$ function, through the estimation of its parameters by nonlinear least squares (NLS).

3.2.2 Structured Shocks

Compared to the solution proposed by Bass et al. (1994) on how to use function $x(t)$, a more practical point of view was proposed by Guseo (2004), who designed some structured shocks for $x(t)$, later successfully employed in the innovation diffusion literature (see, for instance, Guseo and Dalla Valle, 2005, Guseo et al., 2007, Guidolin and Mortarino, 2010, Dalla Valle and Furlan, 2011, Furlan et al., 2016, Bunea et al., 2020). Depending on the type of shock affecting the diffusion process, different forms for $x(t)$ may be designed. In what follows, two main forms of structured shocks are described, the *exponential shock* and the *rectangular shock* (Guseo, 2004), which are easy to include in the model and have proven to be especially effective in capturing external interventions on the diffusion process.

3.2.2.1 Exponential Shock

The exponential shock has been designed to account for intense and fast perturbations that may affect the diffusion dynamics, both positively, such as the acceleration in sales due to marketing strategies and incentive measures, and negatively, such as a dramatic decrease in sales due to the competition of concurrent products, a sudden loss of reputation of the product, or the consequences of an economic crisis. Let I_A be an indicator function taking value 1 if A is true and 0 otherwise. The exponential shock is

$$x(t) = 1 + c_1 e^{b_1(t-a_1)} I_{t \geq a_1}. \tag{3.3}$$

As may be seen in Equation (3.3), function $x(t)$ is centered around the value 1, and a perturbation of exponential form is added. Parameters a_1, b_1, and c_1 have been designed to have a clear and meaningful interpretation: parameter a_1 is the

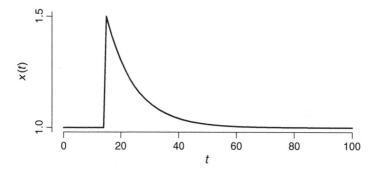

Figure 3.1 Exponential shock with $a_1 = 15$, $b_1 = -0.1$ and $c_1 = 0.5$.

Table 3.1 Generalized Bass model with one exponential shock: parameters and description.

Parameter	Description
m	Market potential
p	Innovation
q	Imitation
a_1	Start of shock
b_1	Memory of shock
c_1	Intensity of shock

starting time of the shock; parameter b_1 measures the "memory" of the shock and is typically negative, suggesting an exponentially decaying behavior; parameter c_1 is the *intensity* of the shock, which may be either positive or negative. The indicator function I takes value 1 when $t \geq a_1$: this suggests that the starting point of the shock needs to be within the observed time window ending with t. Figure 3.1 provides a graphical representation of the shock function, with parameters $a_1 = 15$, $b_1 = -0.1$, and $c_1 = 0.5$. As may be seen in Figure 3.1, $x(t) = 1$ until $a_1 = 15$ when a positive shock of intensity $c_1 = 0.5$ arises, with an exponentially decaying behavior, $b_1 = -0.1$. The exponentially decaying behavior implies that the shock is absorbed with a certain speed, and $x(t)$ returns to its normal behavior, that is, $x(t) = 1$ (Table 3.1). Table 3.1 describes the parameters of a Generalized Bass model with one exponential shock.

3.2.2.2 Rectangular Shock
Rather than being fast and intense, the perturbation may sometimes have a more stable behavior over time, within a limited time span. In this case, the rectangular

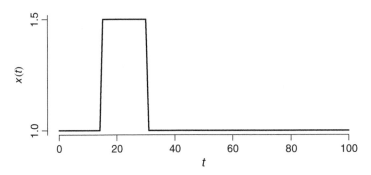

Figure 3.2 Rectangular shock with $a_1 = 15$, $b_1 = 30$ and $c_1 = 0.5$.

shock may be a suitable choice. The rectangular shock is defined as follows

$$x(t) = 1 + c_1 I_{a_1 \leq t \leq b_1}. \tag{3.4}$$

Equation (3.4) shows that, once again, the function $x(t)$ is centered around the value 1, and a rectangular perturbation is added. Parameters a_1 and b_1 indicate the starting and ending times of the shock, respectively, whereas parameter c_1 represents the intensity of such perturbation, which may be either positive or negative.

The rectangular shock is especially suitable whenever a perturbation to the diffusion process occurs in a longer period of time as a result of advertising campaigns or some policies or regulations, within a defined time window. Figure 3.2 provides a graphical representation of it, with parameters $a_1 = 15$, $b_1 = 30$, and $c_1 = 0.5$. In this case, a positive perturbation of intensity $c_1 = 0.5$ starts in $a_1 = 15$ and ends in $b_1 = 30$. After that, the function $x(t)$ returns to be equal to 1 (Table 3.2). Table 3.2 describes the parameters of a Generalized Bass model with one rectangular shock.

Table 3.2 Generalized Bass model with one rectangular shock: parameters and description.

Parameter	Description
m	Market potential
p	Innovation
q	Imitation
a_1	Start of shock
b_1	End of shock
c_1	Intensity of shock

3.2.2.3 More Complex Shocks

One of the great advantages of the use of such structured functions is the possibility to combine them, in order to obtain more complex $x(t)$ functions, displaying their effects at various points in time. A straightforward extension of the above-described perturbations is the setup of an $x(t)$ function with two exponential shocks, namely

$$x(t) = 1 + c_1 e^{b_1(t-a_1)} I_{t \geq a_1} + c_2 e^{b_2(t-a_2)} I_{t \geq a_2},$$

or with two rectangular shocks

$$x(t) = 1 + c_1 I_{t \leq a_1} I_{t \leq b_1} + c_2 I_{t \leq a_2} I_{t \leq b_2}.$$

Another useful possibility involves the combination of an exponential and a rectangular shock, giving rise to the mixed structure

$$x(t) = 1 + c_1 e^{b_1(t-a_1)} I_{t \geq a_1} + c_2 I_{t \leq a_2} I_{t \leq b_2}.$$

Although in theory structured shocks with more than two perturbing functions can be created, practical experience has shown that a too-complex $x(t)$ function is poorly estimated, moreover endangering model parsimony, which is one of the great merits of the BM and GBM.

3.3 Generalized Bass Model: Case Studies

This section is dedicated to the practical application of the GBM to one example already illustrated for the BM, namely, Apple iPhone. Moreover, a new application is discussed, by employing the data of another Apple product, the Mac computer.

3.3.1 Model Fit

The modeling strategy essentially follows what has already been done for the BM. Taking advantage of the closed-form solution, the model is therefore applied to cumulative data, as follows:

$$w(t) = m \, \frac{1 - e^{-(p+q) \int_0^t x(\tau)d\tau}}{1 + \frac{q}{p} e^{-(p+q) \int_0^t x(\tau)d\tau}} + \varepsilon(t),$$

where $\varepsilon(t)$ is an error term and a suitable specification of $x(t)$ is needed. Depending on the selected structure of $x(t)$, the NLS algorithm requires some starting values for the parameters.

As a general indication, m, p, and q may be set equal to those estimated with the BM, because the BM is nested in the GBM, and its estimated parameters may be considered a reliable starting point for NLS estimation of the GBM, whereas the parameters concerning the shock may be based on a hypothesis supported by a careful inspection of the observed data. Through a graphical analysis of data, one may make some hypotheses on the presence of a particular shock and try to estimate it. Sometimes, the knowledge of external events that may have affected the product life cycle may aid this data-driven hypothesis.

3.3.2 Apple iPhone

The first application of the GBM refers to the Apple iPhone, already modeled with the BM. A GBM with one exponential shock, GBMe1, is applied to the series, and the results are reported in Figure 3.3 and Table 3.3. Overall, the model obtains a satisfactory performance, $R^2 = 0.9997$, and all the parameters are statistically significant, as confirmed by the analysis of confidence intervals. This provides a first important confirmation of the model proposed. The statistical significance of the parameters concerning the shock is a clear indication of the relevance of that shock within the diffusion. In this case, the shock has been estimated to start around Q3'10, given that $a_1 = 12.5$ (see the Figure 3.3a and Table 3.3), with a positive intensity of $c_1 = 1.13$ and a negative memory of $b_1 = -0.14$, indicating the tendency of the system to "forget" the effects of the shock and return to the

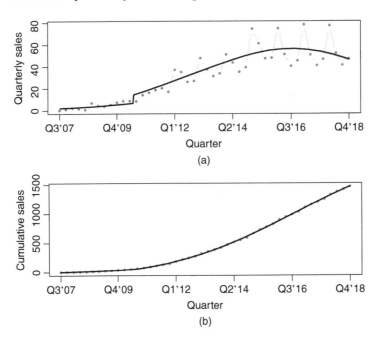

Figure 3.3 GBM with one exponential shock for iPhone. Quarterly sales (a) and cumulative sales (b).

Table 3.3 Parameter estimates of GBMe1 for iPhone.

Parameter	Estimate	s.e.	Lower c.i.	Upper c.i.
m	2108.9	124.9	1864.1	2354.8
p	0.0009	0.00009	0.0007	0.0011
q	0.10	0.011	0.08	0.12
a_1	12.50	0.99	10.56	14.44
b_1	−0.14	0.06	−0.25	−0.03
c_1	1.13	0.17	0.78	1.47

$R^2 = 0.9997$.

normal behavior described by the diffusion parameters p and q. The introduction of a suitable shock allows for providing a description of diffusion patterns that depart from the classical bell-shaped curve of the BM. In the iPhone case, the data show an acceleration of the diffusion process starting around the end of 2010. This positive perturbation to sales may be imputable to an intense advertising campaign exerted by Apple.

Concerning the parameters m, p, and q, the very low value of the innovative component, $p = 0.0009$, testifies that the diffusion process had a slow start, probably because a limited number of pioneer consumers were ready to adopt the new technology when launched into the market and Apple was not a market leader in telephony at that time. This appears coherent with the estimated shock, suggesting that Apple had to intervene to stimulate an otherwise slow process. The value of the estimated market potential, $m = 2108.9$, is higher than that obtained with the BM ($m = 1824$). This may be appreciated in Figure 3.4, where a comparison between the BM and GBMe1 is displayed. As may be observed, the trajectory of the GBMe1 predicts a longer life cycle for iPhone with respect to the BM, consistently with a higher estimate of the market potential m. Indeed, the predictions performed with the GBMe1 appear to better follow the mean behavior of data, especially in early stages of the process and around the last available observations. A further confirmation about the suitability of the GBMe1 comes from the calculated $\tilde{R}^2$ index and F ratio, that is, $\tilde{R}^2 = (0.9997 - 0.9995)/(1 - 0.9995) = 0.4$ and $F = 0.4(46 - 6)/(1 - 0.4)3 = 8.88$.

3.3.3 Apple Mac

The second application of the GBM considers the sales of another technology produced by Apple, namely, the Mac computer. As in the iPhone case, the data are publicly available on www.apple.com. Figure 3.5 displays quarterly sales data (in

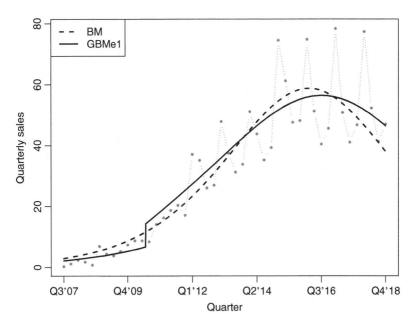

Figure 3.4 Model comparison. GBM with one exponential shock and BM for iPhone.

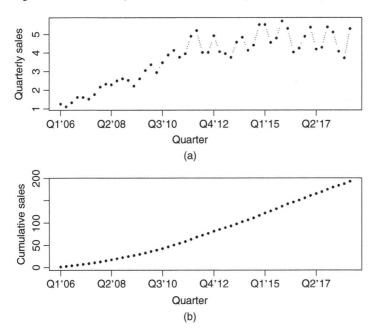

Figure 3.5 Mac. Quarterly (a) and cumulative (b) sales from 2006 to 2018 (million units).

million units), from the first quarter of 2016, Q1'06, to the fourth quarter of 2018, Q4'18, and the corresponding cumulative series. As in the case of other Apple products, the series is characterized by a clear nonlinear trend and a seasonal component, which is more evident from 2009 onwards. Moreover, the data show a more stable behavior from the second half of 2012. In what follows, we propose a direct estimation of a GBM, and we discuss the application of a BM for comparison purposes only, at a second stage. For this example, a GBM with one rectangular shock, GBMr1, is selected, the results of which are presented in Figure 3.6 and Table 3.4. Overall, the model obtains a satisfactory performance, $R^2 = 0.9999$, and all the parameters are statistically significant. In particular, the introduction of the rectangular shock has been particularly effective in capturing the acceleration in the diffusion pattern starting in 2009, likely due to technological updates and new releases that increased sales. Indeed, a positive shock, $c_1 = 0.15$, is estimated to start in Q4'09, $a_1 = 14.7$, and end in Q2'12, $b_1 = 25.7$. A comparison between the predicted trajectories of the GBMr1 and the BM, displayed in Figure 3.7, demonstrates the improvement brought by the GBMr1 with respect to the BM. As already noticed in the case of the iPhone, the inclusion of the shock implies a better description of the entire process with respect to the BM, especially around the first and last available data points. Also in this case, the significant

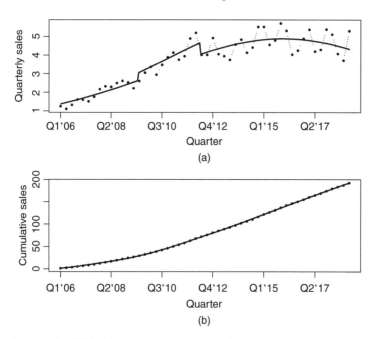

Figure 3.6 GBM with one rectangular shock for Mac. Quarterly sales (a) and cumulative sales (b).

Table 3.4 Parameter estimates of GBMr1 for Mac.

Parameter	Estimate	s.e.	Lower c.i.	Upper c.i.
m	298.9	5.17	288.8	309.1
p	0.0044	0.00007	0.0042	0.0045
q	0.056	0.001	0.054	0.058
a_1	14.7	0.94	12.9	16.6
b_1	25.7	0.71	24.3	27.1
c_1	0.15	0.02	0.12	0.19

$R^2 = 0.99993$.

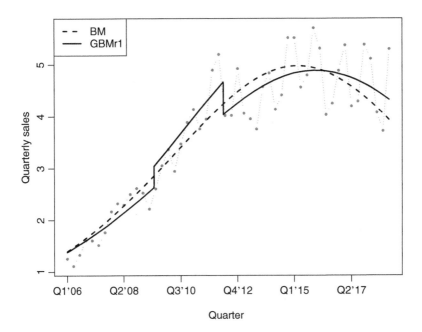

Figure 3.7 Model comparison. GBM with one rectangular shock and BM for Mac.

improvement of the GBMr1 with respect to the BM may be calculated, $\tilde{R}^2 = 0.71$ and $F = 37.5$, providing a further evidence in favor of the model selected.

The application may be further extended by considering the introduction of a second shock, of exponential nature, so the resulting model is a GBM with mixed shocks, one exponential and one rectangular, GBMe1r1. The results of this application are displayed in Table 3.5 and Figure 3.8. The exponential shock is estimated to be positive, $c_1 = 0.35$, starting in Q3'07, $a_1 = 6.86$, and with negative memory,

Table 3.5 Parameter estimates of GBMe1r1 for Mac.

Parameter	Estimate	s.e.	Lower c.i.	Upper c.i.
m	298.1	4.55	289.2	307.0
p	0.0039	0.0001	0.0036	0.0042
q	0.057	0.001	0.055	0.059
a_1	6.86	0.89	5.11	8.61
b_1	−0.24	0.13	−0.51	0.02
c_1	0.35	0.15	0.057	0.65
a_2	17.6	0.83	15.9	19.2
b_2	25.4	0.52	24.4	26.5
c_2	0.19	0.02	0.14	0.24

$R^2 = 0.99995$.

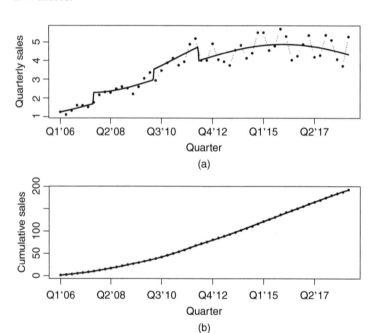

Figure 3.8 GBM with one exponential and one rectangular shock for Mac. Quarterly sales (a) and cumulative sales (b).

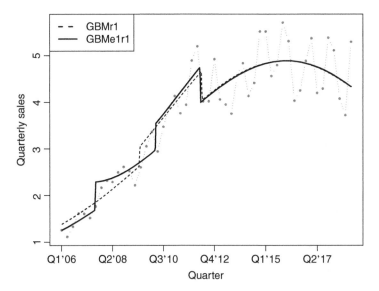

Figure 3.9 Model comparison. GBM with one exponential and one rectangular shock and GBM with one rectangular shock for Mac.

$b_1 = -0.24$. The inclusion of this shock implies a reduction of the time window for the rectangular shock, which is now estimated to start in Q4'09, $a_2 = 17.6$. Overall, the improvement with respect to the GBMr1 is not particularly strong, since $\tilde{R}^2 = 0.28$ and $F = 5.7$. However, this application provides an instructive example for the use of a mixed shock. Figure 3.9 provides a graphical comparison between GBM with one exponential and one rectangular shock and GBM with one rectangular shock for Mac.

3.4 Recap

At the end of the chapter, some important properties and limitations of the GBM are summarized. Similarly to the BM, the GBM assumes a finite life cycle for a product, but thanks to the inclusion of structured shocks it may describe diffusion patterns that depart from the (instantaneous) bell-shaped curve of the BM. This is especially useful when suitable forms of $x(t)$ may be employed.

Properties

- The GBM allows one to account for structured shocks that may affect the timing of the diffusion process through a simple intervention function $x(t)$.

- The GBM has a closed-form solution that makes its use very straightforward with only aggregate data on sales or adoptions.
- The parameters of the GBM are easy to interpret and provide a useful explanation to life cycle perturbations, due to market-related events or economic changes.

Limitations

- The GBM may prove to be an extremely flexible model, and the inclusion of several shocks may lead to the risk of overfitting. This calls for a special care in managing the function $x(t)$, and a use of the GBM possibly combined with careful market considerations.
- The GBM is able to identify the shocks in the diffusion process and an ex post confirmation of their effectiveness. Therefore, its use has more of a historical rather than a predictive purpose.

References

Frank M Bass. Empirical generalizations and marketing science: A personal view. *Marketing Science*, 14(3 supplement):G6–G19, 1995.

Frank M Bass, Trichy V Krishnan, and Dipak C Jain. Why the Bass model fits without decision variables. *Marketing Science*, 13(3):203–223, 1994.

Anita M Bunea, Pompeo Della Posta, Mariangela Guidolin, and Piero Manfredi. What do adoption patterns of solar panels observed so far tell about governments' incentive? Insights from diffusion models. *Technological Forecasting and Social Change*, 160:120240, 2020.

Alessandra Dalla Valle and Claudia Furlan. Forecasting accuracy of wind power technology diffusion models across countries. *International Journal of Forecasting*, 27(2):592–601, 2011.

Claudia Furlan, Mariangela Guidolin, and Renato Guseo. Has the Fukushima accident influenced short-term consumption in the evolution of nuclear energy? An analysis of the world and seven leading countries. *Technological Forecasting and Social Change*, 107:37–49, 2016.

Mariangela Guidolin and Cinzia Mortarino. Cross-country diffusion of photovoltaic systems: Modelling choices and forecasts for national adoption patterns. *Technological Forecasting and Social Change*, 77(2):279–296, 2010.

Renato Guseo. Strategic interventions and competitive aspects in innovation life cycle. Technical report, Department of Statistical Sciences, University of Padua, 2004.

Renato Guseo and Alessandra Dalla Valle. Oil and gas depletion: Diffusion models and forecasting under strategic intervention. *Statistical Methods and Applications*, 14(3):375–387, 2005.

Renato Guseo, Alessandra Dalla Valle, and Mariangela Guidolin. World Oil Depletion Models: Price effects compared with strategic or technological interventions. *Technological Forecasting and Social Change*, 74(4):452–469, 2007.

Vijay Mahajan, Eitan Muller, and Frank M Bass. New product diffusion models in marketing: A review and directions for research. *Journal of Marketing*, 54(1):1–26, 1990.

John A Norton and Frank M Bass. A diffusion theory model of adoption and substitution for successive generations of high-technology products. *Management Science*, 33(9):1069–1086, 1987.

Renana Peres, Eitan Muller, and Vijay Mahajan. Innovation diffusion and new product growth models: A critical review and research directions. *International Journal of Research in Marketing*, 27(2):91–106, 2010.

4

Innovation Diffusion with a Dynamic Market Potential: The GGM

After reading this chapter, you should be able to

- Know the basic features of the GGM
- Interpret the parameters of the GGM
- Make predictions with the GGM.

4.1 Introduction

One of the characterizing assumptions of the Bass model (BM) relates to the size of the market potential m, whose value is assumed to be determined at the time of introducing the new product and remains constant along the entire diffusion process. This implies that when an innovation enters a new market, a set of potential consumers are ready to buy it. Such an assumption clearly simplifies the structure of the BM, whose merits are also due to its simplicity. However, Mahajan et al. (1990) observed that theoretically, no rationale exists for a constant population of adopters and a dynamic construct appears as a reasonable choice in several situations. The possibility of a dynamic potential has been addressed in the literature since the 1970s. A detailed review of this literature is presented in Guseo and Guidolin (2009).

From a practical perspective, the assumption of a constant market potential may be reasonable, especially in some circumstances: if a product is a successive generation of something already existing in the market and consumers have already experienced it, then one may assume that a set of potential adopters is present as soon as the novelty arrives. On the contrary, products or technologies with a higher degree of innovativeness will be more difficult to be adopted. This is

Innovation Diffusion Models: Theory and Practice, First Edition. Mariangela Guidolin.
© 2024 John Wiley & Sons Ltd. Published 2024 by John Wiley & Sons Ltd.
Companion Website: www.wiley.com/go/innovationdiffusionmodels/

because of their complexity and the lack of similar products already existing in the market that may be tested and observed. Many new products are characterized by the so-called *incubation* period when whether an innovation will be a success or a failure is still unclear. In this phase, marketing strategies based on advertising and communication are vital to informing about the existence, and the main characteristics of the product in order to stimulate its trial and build the market. In this sense, the market potential may be considered a dynamic entity that develops in time. Otherwise stated, one becomes a potential adopter if and when he/she has received enough information about the product, i.e. its existence and its main characteristics. Recognizing the relevance of communication also justifies the increasing attempt from firms to manage it, not only by planning effective advertising campaigns, but also trying to take advantage of the power of word-of-mouth: in what is sometimes called *viral marketing*, companies are currently investing much effort to take advantage of the relationships between consumers, given that a recommendation from a friend or other trusted source has more credibility than advertisements. So communication about a product is not only advertising, but is also based on information sharing between consumers. Starting from these considerations, Guseo and Guidolin (2009) proposed a model where the market potential depends on the diffusion of information about the product.

The chapter is organized as follows. Section 4.2 introduces a generalization of the BM with a dynamic market potential, $m(t)$. Section 4.3 describes the Guseo–Guidolin model, GGM, i.e. a specification for the general model proposed in Section 4.2. Section 4.4 proposes some generalizations of the GGM, while Section 4.6 deals with the practical application of the GGM to three real-data examples.

4.2 Dynamic Market Potential: Theory

Following Guseo (2004) and Guseo and Guidolin (2009), a generalization of the BM, considering a dynamic market potential, $m(t)$ may be formulated as

$$z'(t) = m(t) \left[p + q \frac{z(t)}{m(t)} \right] \left[1 - \frac{z(t)}{m(t)} \right] + m'(t) \frac{z(t)}{m(t)}, \quad t > 0. \tag{4.1}$$

Equation (4.1) describes instantaneous adoptions $z'(t)$ with a BM with $m(t)$ plus a factor $m'(t)z(t)/m(t)$, which assigns to $z'(t)$ a portion of the variation of the market potential $m'(t)$, namely the growth rate $z(t)/m(t)$. In Equation (4.1), the variation of the market potential $m'(t)$ has an effect on instantaneous adoptions $z'(t)$, which may be either positive and reinforcing, if $m(t)$ is increasing, or negative, if $m(t)$ is decreasing: this expresses the idea that a product's adoptions gain an extra benefit from expanding market potential, whereas a declining market will weaken the process. Equation (4.1) may be conveniently rearranged as follows

$$\frac{z'(t)m(t) - z(t)m'(t)}{m^2(t)} = \left[\frac{z(t)}{m(t)}\right]' = \left[p + q\frac{z(t)}{m(t)}\right]\left[1 - \frac{z(t)}{m(t)}\right]. \tag{4.2}$$

Equation (4.2) provides a useful representation, which, by setting $y(t) = z(t)/m(t)$, yields

$$y'(t) = p + qy(t)(1 - y(t)),$$

which is the standard BM. It is interesting to observe that the solution to Equation (4.1) does not depend on the specification of $m(t)$.

4.2.1 Closed-Form Solution

The generalization of the BM where $m(t)$ is time dependent has closed-form solution

$$z(t) = m(t)\frac{1 - e^{-(p+q)t}}{1 + \frac{q}{p}e^{-(p+q)t}}. \tag{4.3}$$

Equation (4.3) well shows that $m(t)$ is a "free" function, which multiplies the dynamics of the diffusion process expressed by parameters p and q. Function $m(t)$ may take several structures, depending on the hypotheses made about the development of the market potential. In Guseo and Guidolin (2009, 2010) some specific structures for $m(t)$ have been proposed. In the first case, $m(t)$ is designed to describe a communication process that builds the market, in the second to account for the possible presence of network externalities.

4.3 GGM

Guseo and Guidolin (2009) proposed a particular specification for $m(t)$, by making the hypothesis that the development of the market potential depends on a communication process about the new product, which typically precedes the adoption phase and serves the purpose of "building" the market.

In particular, the dynamic market potential $m(t)$ is defined according to a BM-like structure

$$m(t) = K\sqrt{\frac{1 - e^{-(p_c+q_c)t}}{1 + \frac{q_c}{p_c}e^{-(p_c+q_c)t}}}. \tag{4.4}$$

In Equation (4.4), the parameters p_c and q_c govern the communication process (the square root is a technical aspect described in Section 4.3.1). The parameter p_c describes the behavior of innovative consumers who start to talk about the

Table 4.1 GGM: parameters and description.

Parameter	Description
K	Market potential
p_c	Innovation in communication
q_c	Imitation in communication
p_s	Innovation in adoption
q_s	Imitation in adoption

new product, whereas q_c represents those forces spreading the word once they have received the information, making it "viral." The parameter K indicates the asymptotic behavior of $m(t)$ when all informed consumers will eventually become adopters.

The GGM has the following cumulative structure

$$z(t) = K \sqrt{\frac{1 - e^{-(p_c + q_c)t}}{1 + \frac{q_c}{p_c} e^{-(p_c + q_c)t}} \frac{1 - e^{-(p_s + q_s)t}}{1 + \frac{q_s}{p_s} e^{-(p_s + q_s)t}}}, \quad t > 0. \tag{4.5}$$

In Equation (4.5), cumulative adoptions, $z(t)$, are described as the product of two separate and distinct phases, the communication, with parameters p_c and q_c, and the adoption process, with parameters p_s and q_s. Note that the BM may be seen as a special case of GGM where the spread of information is so fast that a set of potential adopters are ready to purchase as soon as the product enters the market, $m(t) = K$. Table 4.1 summarizes the parameters of the model.

Figure 4.1 describes three diffusion processes that differ for the values of p_c and q_c, with p_s and q_s being equal. This well shows that depending on the strength of the communication, expressed by the value of p_c and q_c, the entire diffusion process may be faster or slower. For example, an extremely good communication process, (e.g. $p_c = 0.01$ and $q_c = 0.9$), will lead to a fast generation of the market potential, with positive effects on the entire process. On the other hand, weak communication, (e.g. $p_c = 0.001$ and $q_c = 0.05$), will cause a slow development of the market and will eventually slow down observed sales.

4.3.1 Specification of $m(t)$ in the GGM

As we have seen, in the GGM the market potential $m(t)$ is assumed to depend on the evolution of a communication process about the new product, as in Equation (4.4). The market potential may be seen as a measure of the *absorptive capacity* of a system, which is the ability to recognize the value of new information, assimilate it, and use it (Cohen and Levinthal, 1990). This ability is a

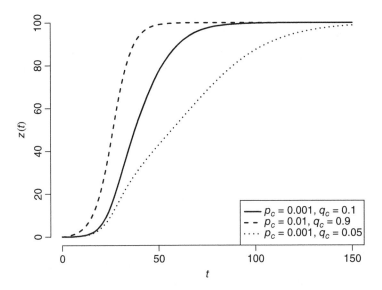

Figure 4.1 Three cumulative diffusion processes with different combinations of p_c and q_c parameters. Parameters $p_s = 0.001$ and $q_s = 0.2$. The asymptotic market potential K is fixed at $K = 100$.

function of prior knowledge: at the individual level, this is related to the cognitive functions of the single person, while at the system level, this has to do with the development of common knowledge through communication and information sharing. A process of communication and information sharing may be thought of as a *network*, in which individuals connect to each other and share information, in our case, about the product.

To make this point more intuitive, we may use the typical concepts of network science (e.g. Newman, 2018). Let $G = (V, E)$ be a graph, with $V = \{1, \dots, N\}$ the set of *nodes* (individuals), with cardinality $N = c(V)$. The set E of *edges* is the set of ordered pairs (i, j), i.e. $E \subseteq \{(i, j)\} : i, j \in N, E \subset V \times V$, and describes all the possible binary relationships between the nodes V (including self-loops). So the number of edges is, as order of magnitude, the square of the number of nodes.

A network is typically represented through an *adjacency matrix* Y, which is a square matrix of size $N \times N$, where $Y_{i,j} = 1$ if there has been a connection in the pair (i, j), otherwise $Y_{i,j} = 0$. For example, a simple 3×3 adjacency matrix is

$$Y = \begin{pmatrix} 0 & 1 & 1 \\ 1 & 0 & 0 \\ 1 & 0 & 0 \end{pmatrix}.$$

In a communication network, $Y_{i,j} = 1$ means that there has been communication between nodes i and j. In principle, under perfect communication, all pairs $Y_{i,j}$

would be equal to 1. However, it is more realistic that not all nodes will be connected, so that some $Y_{i,j} = 0$. This implies that the *size of the network*, i.e. the number of edges E, will be $U \leqslant N \times N$, due to communication constraints.

In the GGM, we assume that the market potential is formed by all individuals aware of the product so that the market potential is the result of a communication process. We also assume that the evolution over time of this communication process, i.e. the number of active edges, follows a Bass model (BM)

$$h(t) = m_c v(t) = m_c \frac{1 - e^{-(p_c + q_c)t}}{1 + \frac{q_c}{p_c} e^{-(p_c + q_c)t}}, \tag{4.6}$$

where coefficients m_c, p_c, and q_c are the parameters of the BM for the communication process; parameters p_c and q_c describe innovative and imitative behavior in information sharing: some connections are created first (innovators), whereas others are established later by imitation, while m_c is the total amount of possible connections in the network.

In Equation (4.6), the quantity $h(t)$ represents the number of active edges of a network, but for the definition of the market potential $m(t)$ we are interested in the number of *informed individuals* (i.e. nodes V). An approximation for the market potential $m(t)$ is thus obtained in a straightforward way, by considering the squared root of $h(t)$

$$\sqrt{h(t)} = \sqrt{m_c} \sqrt{v(t)} = K \sqrt{\frac{1 - e^{-(p_c + q_c)t}}{1 + \frac{q_c}{p_c} e^{-(p_c + q_c)t}}}.$$

4.3.2 Communication and Adoption in the GGM

Sometimes it is useful to express the GGM in a compact way as follows

$$z(t) = K \, S(t; p_c, q_c, p_s, q_s) = K \, \sqrt{F(t; p_c, q_c)} \, G(t; p_s, q_s). \tag{4.7}$$

In Equation (4.7) $S(t; p_c, q_c, p_s, q_s)$ is the product of two c.d.f. $\sqrt{F(t; p_c, q_c)}$ and $G(t; p_s, q_s)$, representing the communication and adoption phases, respectively.

The corresponding instantaneous process $z'(t)$ may be defined accordingly

$$z'(t) = K \, S'(t; p_c, q_c, p_s, q_s) \tag{4.8}$$

with $S'(t) = S'(t; p_c, q_c, p_s, q_s) = \frac{1}{2} F(t)^{-\frac{1}{2}} G(t) f(t) + \sqrt{F(t)} g(t)$, where $f(t) = F'(t)$ and $g(t) = G'(t)$. In particular, we may use a more compact notation for $S'(t)$, to highlight the presence of two phases in the diffusion process

$$S'(t) = \frac{1}{2} F(t)^{-\frac{1}{2}} G(t) f(t) + \sqrt{F(t)} g(t) = k_1(t) + k_2(t). \tag{4.9}$$

Functions $k_1(t)$ and $k_2(t)$ essentially describe communication and adoption phases, respectively. In Guseo and Guidolin (2011), it is shown that, based on

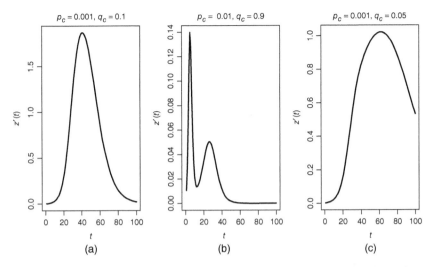

Figure 4.2 Three instantaneous diffusion processes with different combinations of p_c and q_c parameters. Parameters $p_s = 0.001$ and $q_s = 0.2$. (a) $p_c = 0.001$, $q_c = 0.1$, (b) $p_c = 0.01$, $q_c = 0.9$, (c) $p_c = 0.001$, $q_c = 0.05$.

the temporal positioning of these two phases, which depends on the values of parameters p_c, q_c, p_s, and q_s, instantaneous adoptions in the GGM may have a unimodal or bimodal pattern. As shown in Figure 4.2, different values of communication parameters p_c and q_c, cause the GGM to have very different forms.

Also, based on the values of the parameters p_c, q_c, p_s, and q_s, it is possible to define the mode t^+, the median $t_{0.5}$, and the mean $\bar{t}$ for both phases, as follows

$$t^+_{com} = \frac{\log(q_c/p_c)}{p_c + q_c}, \quad {}_{com}t_{0.5} = \frac{1}{p_c + q_c} \log\left(2 + \frac{q_c}{p_c}\right), \quad \bar{t}_{com} = \frac{1}{q_c} \log\left(1 + \frac{q_c}{p_c}\right),$$

$$t^+_{ado} = \frac{\log(q_s/p_s)}{p_s + q_s}, \quad {}_{ado}t_{0.5} = \frac{1}{p_s + q_s} \log\left(2 + \frac{q_s}{p_s}\right), \quad \bar{t}_{ado} = \frac{1}{q_s} \log\left(1 + \frac{q_s}{p_s}\right).$$

The definition of these location indexes allows us to correctly place the communication and adoption phases in time. In general, the communication phase precedes the adoption one; however, in some peculiar cases, the reverse may occur, and the adoption phase "comes before" communication. This may happen in those cases where the new product is highly expected from consumers, and there is an accumulation of demand prior to the product launch so that as soon as the product enters the market there is a group of consumers ready to buy it. In these cases, communication has the role of strengthening and stimulating the diffusion process, after the initial boost of early adopters (Guseo and Guidolin, 2011).

4.4 Generalizations of the GGM

This section presents some generalizations of the GGM, by accounting for external perturbations, both on the communication and adoption phases.

4.4.1 GGM with Structured Shocks

A possible generalization of the GGM has been illustrated in Furlan et al. (2016) by introducing a GGM with structured shocks, where a function $x(t)$ perturbs the adoption phase, in a Generalized Bass model (GBM)-like fashion, i.e.

$$z(t) = K \sqrt{\frac{1 - e^{-(p_c+q_c)t}}{1 + \frac{q_c}{p_c} e^{-(p_c+q_c)t}} \frac{1 - e^{-(p_s+q_s)\int_0^t x(\tau)d\tau}}{1 + \frac{q_s}{p_s} e^{-(p_s+q_s)\int_0^t x(\tau)d\tau}}}, \quad t > 0. \tag{4.10}$$

In Furlan et al. (2016), model (4.10) has been employed by using a rectangular shock as described in Equation (3.4), thus terming the final model GGM-R.

Interestingly, the GBM and the BM can be seen as special cases of model (4.10), when $m(t) = K$, and $x(t) = 1$, and $m(t) = K$, respectively (Furlan et al., 2016). Also, model (4.10) may be further generalized by including an intervention function in the communication phase. A proposal in this sense has been made by Guidolin and Guseo (2014), by designing a GGM with shocks to account for seasonality

$$z(t) = K \sqrt{\frac{1 - e^{-(p_c+q_c)\int_0^t x_F(\tau)d\tau}}{1 + \frac{q_c}{p_c} e^{-(p_c+q_c)\int_0^t x_F(\tau)d\tau}} \frac{1 - e^{-(p_s+q_s)\int_0^t x_G(\tau)d\tau}}{1 + \frac{q_s}{p_s} e^{-(p_s+q_s)\int_0^t x_G(\tau)d\tau}}},$$

where $x_F(t)$ and $x_G(t)$ are harmonic functions, i.e. $x_F(t) = 1 + a\cos\left(\frac{2\pi t}{s}\right) + b\sin\left(\frac{2\pi t}{s}\right)$, $x_G(t) = 1 + c\cos\left(\frac{2\pi t}{f}\right) + d\sin\left(\frac{2\pi t}{f}\right)$.

4.4.2 GGM with Deterministic Seasonality

Guidolin and Guseo (2014) also developed a generalization of the GGM to account for a deterministic seasonality, whose effect is stronger around the peak of sales and weaker during product launch and decline. In particular, a GGM with an additive seasonal component proportional to the trend has been defined as follows

$$z'(t) = (K + A(t)) S'(t),$$

where $S'(t)$ is the instantaneous diffusion process obeying a GGM, according to Equation (4.9), while $A(t)$ describes seasonality, whose effect is proportional to

$S'(t)$. The function $A(t)$ may be represented with a linear combination of trigonometric functions such as

$$A(t) = \sum_{j=1}^{\left[\frac{s}{2}\right]} \left[a_j \cos\left(\frac{2\pi jt}{s}\right) + b_j \sin\left(\frac{2\pi jt}{s}\right) \right],$$

where $[s/2]$ is the integer part of $s/2$.

4.5 A Dynamic Market Potential Model with Network Externalities

In Sections 4.1–4.3, it has been suggested that many innovations may experience difficulties in the early stages of their life cycle because limited knowledge about them penalizes sales. The first phase of an innovation diffusion process indicated as the "incubation period," has proven to have a chilling effect on new product growth, and the GGM was aimed at recognizing that effect, whereas the standard BM does not. Though chilling effects exerted by incubation seem typical of many innovations at their first generation, this phenomenon is especially visible in the case of network goods, such as telephony, electricity, Internet, and social networks, whose benefit and adoption by the single individual depends on the perceived number of others who have already adopted and thus on the creation of a physical network of interacting units. The increase of a product's utility for a consumer as the number of other users does is defined as a *network externality*. Network externalities may be direct, when the utility is directly affected by the number of other users, as in the case of social networks or telecommunications, or indirect, when the increase of utility occurs through market mediation, so that, for example, the number of users of DVD players influences the number of DVD titles available in rental outlets. While indirect network externalities require an analysis of market mechanisms, the case of direct externalities specifically deals with those interdependences between consumers that tie together their utilities. The economic theory on network externalities has generally focused on their effect on industry structure, suggesting that network effects provide a competitive advantage for producers with larger market share over later entrants (Shapiro and Varian, 1999): as a consequence, this may lead to a greater tendency toward monopolization and influence optimal policy choices of firms, such as the adoption of low pricing strategies to deter the entry of competitors. Thus, the ability to build network effects has been generally considered an important factor to facilitate the success of a producer in its market. However, in the early stages of the diffusion of innovation with strong network externalities, the rate of adoption proceeds extremely slowly and this extended incubation period is often reflected in the long left tail of the diffusion curve. This suggests that pioneer firms that introduce new network goods into markets do not only benefit from the opportunity to create a network for their technologies as first market entrants but also have to sustain some costs to achieve this position, given that the rate of adoption is initially very low. Thus, network externalities may be both a source of benefits and disadvantages for firms, and analyzing the evolution of the markets for network goods may help to understand both sides of this story. Network industries typically exhibit a positive *critical mass*, which is the minimal number of adopters of an interactive innovation for the further rate of adoption to be self-sustaining. The concept of critical mass formalizes the so-called "chicken and egg" dilemma, for which many consumers are not interested in purchasing the good because the installed base is too small, but the installed base is too small because an insufficient number of consumers have adopted the good. The importance of the critical mass for network goods is related to their specific nature: while many new products or services provide benefits for the single adopter, independently of the purchase decision of others, in the case of network or interactive products network externalities apply, that is, the utility of a single adopter is strongly dependent on what others decide to do since these goods need a network to be used. Most innovations present such a degree of complexity, network effects, and investment riskiness, which the average consumer is initially discouraged to adopt and will eventually decide to adopt when there is a sufficient guarantee of the investment's safety: many consumers will wait until a critical mass is reached and if all consumers acted like this, a critical mass would never be achieved. However, individuals may be characterized by different levels of receptiveness to innovations, so some of them will purchase the innovation before reaching critical mass. These different levels of receptiveness or resistance to innovations represent a primary source of heterogeneity among the components of a population: different individuals will have different thresholds. The concept of individual threshold is defined as the proportion of the group for an individual to engage in a particular behavior. In the innovation diffusion context, an individual threshold is a personal evaluation to be compared with the relative number of other individuals that must have adopted the innovation before a given potential adopter will adopt it, and the case of network or interactive goods appears particularly suitable for being treated with threshold models. Following Guseo and Guidolin (2010), the relationship between network externalities and threshold levels can be explained by considering an individual's utility function. The consumer will purchase the product if the number of adopters is larger than a certain level depending on personal evaluations of the relative advantage of the good.

Guseo and Guidolin (2010) defined a *consumer index* $h = p/w$, where p denotes the public price of a good and w describes the individual *willingness to pay* of a consumer, which is a function of the growth of the product. An individual may become a potential adopter if he/she has an index $h \leq 1$. The heterogeneity of individuals may be represented through a distribution of $h \in H_t$. Individual threshold h is compared with the rate of growth of the product $y(t)$.

This makes it possible to define the market potential at time t with the probability distribution $P(h)$ of H_t, for $h \leq y(t)$,

$$m(t) = KP[H_t \leq y(t)]. \tag{4.11}$$

Distribution $P(h)$ is typically the normal distribution,

$$P(H_t \leq v(t)) = \Phi\left[\frac{v(t) - \mu(t)}{\sigma}\right]. \tag{4.12}$$

The final dynamic market potential model with network externalities is

$$z(t) = K\Phi\left[\frac{v(t) - \mu(t)}{\sigma}\right] \frac{1 - e^{-(p+q)t}}{1 + \frac{q}{p}e^{-(p+q)t}}, \quad t > 0. \tag{4.13}$$

4.6 GGM: Case Studies

In this section, three applications of the GGM are illustrated. The first one is again based on the iPhone data, the second shows the case of the iPod (available at www.apple.com), illustrated in Chapter 1, and a third application concerns sales data of Samsung smartphones (available at www.statista.com). These three cases will be useful to show the flexibility of the GGM.

4.6.1 Model Fit

The fitting of the GGM is based on:

$$w(t) = K\sqrt{\frac{1 - e^{-(p_c+q_c)t}}{1 + \frac{q_c}{p_c}e^{-(p_c+q_c)t}} \frac{1 - e^{-(p_s+q_s)t}}{1 + \frac{q_s}{p_s}e^{-(p_s+q_s)t}}} + \varepsilon(t),$$

where $w(t)$ is the observed series, and $\varepsilon(t)$ is an error term, for which usual assumptions hold.

As in the BM and GBM cases, the application of the GGM is based on the estimation of the model on cumulative data through the nonlinear least squares (NLS) procedure. As already seen in previous applications, also for the GGM the NLS estimation requires the specification of some starting points for the parameters, K, p_c, q_c, p_s, and q_s. Given that the BM may be considered a special case of the GGM, one may consider the estimates of m, p, and q as reliable initial estimates for K, p_s, and q_s. Instead, p_c and q_c require some working hypothesis, and practical experience has shown that a good choice is $p_c = 0.001$ and $q_c = 0.01$. The improvement obtained with the GGM over the BM may be evaluated through the usual indexes, $\tilde{R}^2$ and F, given the BM is nested in the GGM as a special case, where $K = m$.

4.6.2 Apple iPhone

Besides the BM and GBM, the case of the iPhone is also considered to show the implementation of the GGM. The results of this application are displayed in Figure 4.3 and Table 4.2. As a general observation, the model is very well-estimated and all the parameters are statistically significant. Hence, the inclusion of a dynamic market potential, depending on the communication parameters p_c and q_c, appears a very suitable choice for this product. This suggests that the market potential for this technology has been built over time, as a function of a strong diffusion of information, driven by effective advertising campaigns and the spread of information between consumers. Both communication and adoption phases appear to be especially characterized by imitative behavior, (i.e. $q_c = 0.20$ and $q_s = 0.10$), whereas pioneer consumers are, as one could

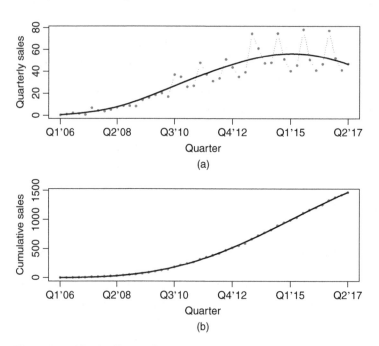

Figure 4.3 GGM for iPhone. Quarterly sales (a) and cumulative sales (b).

Table 4.2 Parameter estimates of GGM for iPhone.

Parameter	Estimate	s.e.	Lower c.i.	Upper c.i.
K	2116.7	97.5	1925.7	2307.9
p_c	0.006	0.002	0.003	0.009
q_c	0.20	0.04	0.13	0.28
p_s	0.002	0.0003	0.002	0.003
q_s	0.10	0.007	0.09	0.11

$R^2 = 0.9997$.

expect, a minor proportion in both phases, $p_c = 0.006$ and $p_s = 0.002$. The fit illustrated in Figure 4.3 highlights the extremely good performance of the model in describing the mean trajectory of the series, and the comparison with the BM proposed in Figure 4.4 well shows the merits of the GGM. One may note that the GGM is more effective in capturing the structure of data at the beginning and at the end of the observed series: In this sense, the GGM is able to "correct" a typical flaw of the BM, namely the tendency of overestimating the first part of the life cycle and underestimating the last part of it. For this reason, the GGM seems to predict a longer life cycle for the iPhone with respect to the BM, as may be seen in Figure 4.4, and by comparing the estimated market potentials for the GGM and BM, respectively, $K = 2116.7$ and $m = 1823.7$. Concerning the improvement obtained with the GGM, in this case we obtain $\tilde{R}^2 = 0.4$ and $F = 13.3$, which confirms the significant gain.

Since the iPhone data have been used as an example for both the GBM and the GGM, one could wonder what is the right model to select. Indeed, both models have satisfactory performance and provide convincing modeling. A model selection between GBM and GGM through $\tilde{R}^2$ and F is not possible, as they are not nested models, and the direct comparison of the R^2 would not be correct, as the GBM has a higher one just for having more parameters. Hence, the choice should be based on other considerations, referring to the market and the characteristics of the product being studied. In this case, the exponential shock was explained as the effect of a well-developed advertising and communication campaign, while the GGM models the iPhone life cycle as the result of two separate phases, communication and adoption. In both cases, communication appears to have had a crucial role in stimulating the diffusion process, and both models provide a reasonable explanation for the behavior of data. However, this is rarely the case: in many other situations, one of the two models, GBM or GGM, is not convincing enough, and the "dilemma" of choosing between the two options is not posed.

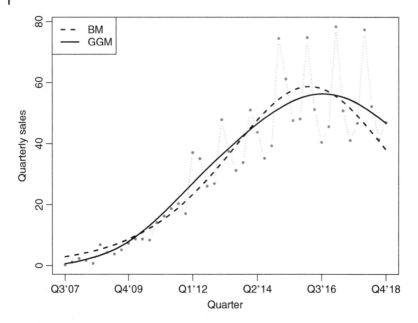

Figure 4.4   Model comparison. GGM and BM for iPhone.

4.6.3   Apple iPod

The second example, concerning the life cycle of the Apple iPod, well illustrates the flexibility of the GGM in describing the mean behavior of the series. Parameter estimates are shown in Table 4.3. Figure 4.5 reports the predictions according to the model. The inspection of quarterly data in Figure 4.5a is especially suggestive, showing that the GGM can describe in an efficient way both the first and last portions of observed data, and also performs a "plateauing"

Table 4.3   Parameter estimates of GGM for iPod.

Parameter	Estimate	s.e.	Lower c.i.	Upper c.i.
K	4175.5	29.7	4117.3	4233.7
p_c	0.0005	0.0001	0.0004	0.0007
q_c	0.15	0.004	0.14	0.16
p_s	0.0014	0.0003	0.0008	0.0020
q_s	0.30	0.019	0.26	0.34

$R^2 = 0.9997$.

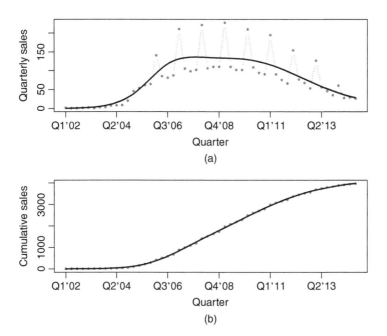

Figure 4.5 GGM for iPod. Quarterly sales (a) and cumulative sales (b).

prediction of the maturity phase of the life cycle, which appears especially suitable for this product. More in detail, one could observe that the model appears to predict an almost bimodal behavior, with a first peak around Q4'06 and a kind of saddle and a second less evident peak at the end of the year 2010.

The possibility of this bimodal behavior has been studied in detail by Guseo and Guidolin (2011), providing a possible explanation for the presence of a slowdown in some diffusion processes, a phenomenon also termed "chasm" in the marketing literature (Moore, 1999). This phenomenon refers to the behavior of some innovative products that, after reaching a local peak, experience a phase of slowdown in sales, followed by the second phase of growth, which leads to reaching the global maximum peak. The comparison with the prediction obtained with a simple BM, shown in Figure 4.6, well clarifies the difference between the two models in this particular example. It may be observed that the GGM outperforms the BM, especially in the first phase of the life cycle, until the end of the year 2004, where the BM visibly overestimates the observed series. On the other hand, the BM provides a good fit in the last portion of data, not so different from that of the GGM: this is because the BM has a satisfactory performance when the life cycle is essentially complete.

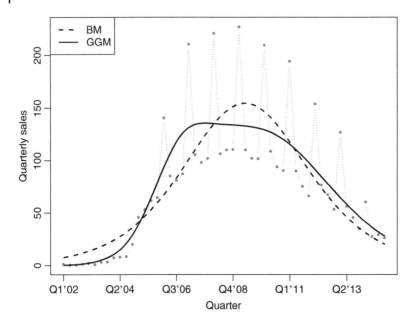

Figure 4.6 Model comparison. GGM and BM for iPod.

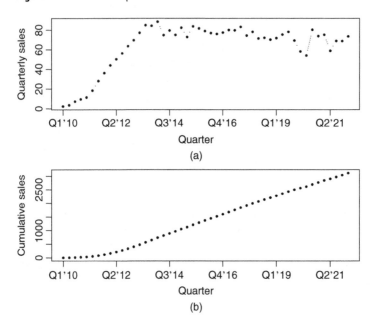

Figure 4.7 Samsung smartphones. Quarterly (a) and cumulative (b) sales from 2010 to 2022 (million units).

4.6.4 Samsung Smartphones

The third application concerns sales of Samsung smartphones, whose series goes from Q1'10 to Q4'21. The usual display of quarterly and cumulative data in Figure 4.7 highlights that quarterly sales had a strongly asymmetric growth, with a fast growth until the end of 2012 and a slow decrease starting from 2013. As in the case of the iPod, the series exhibits a "plateauing" pattern, which surely departs from the bell-shaped behavior of the BM. In order to describe this series and possibly account for its asymmetric nature, a GGM is fit to data. The estimation results presented in Table 4.4 and Figure 4.8 confirm the selection of the GGM as

Table 4.4 Parameter estimates of GGM for Samsung smartphones.

Parameter	Estimate	s.e.	Lower c.i.	Upper c.i.
K	4030.7	75.47	3882.8	4178.6
p_c	0.0015	0.00001	0.0014	0.0016
q_c	0.08	0.0026	0.08	0.09
p_s	0.012	0.0006	0.011	0.014
q_s	0.21	0.008	0.20	0.23

$R^2 = 0.9999$.

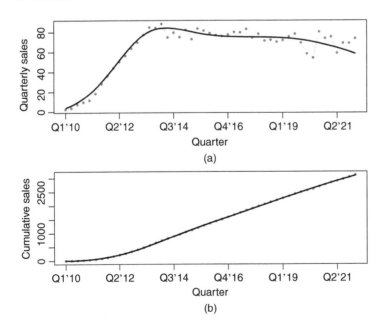

Figure 4.8 GGM for Samsung smartphones. Quarterly sales (a) and cumulative sales (b).

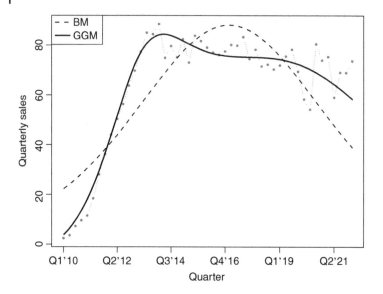

Figure 4.9 Model comparison. GGM and BM for Samsung smartphones.

a suitable model. In terms of estimates, one may observe that the adoption phase has been characterized by higher parameters, $p_s = 0.012, q_s = 0.21$, with respect to the corresponding communication phase, $p_c = 0.0015, q_c = 0.08$.

The most interesting insight comes from the comparison with the BM, displayed in Figure 4.9, where it is highly evident the much better fit of the GGM. In particular, the GGM is able to capture the presence of a slowdown in sales, reasonably imputable to the presence of the two phases within the life cycle, i.e. communication and adoption.

4.7 Recap

We conclude this chapter by summarizing some important properties of the GGM, along with some limitations.

Properties

- The GGM extends the basic structure of the BM, by accounting for a time-dependent market potential $m(t)$. Note that the choice for $m(t)$ in the GGM is a special one, but $m(t)$ may have many other possible forms, depending on modeling needs

- The GGM is a very flexible model that can describe both unimodal and bimodal structures in data.
- In the GGM, the dynamic market potential is dependent on a communication process, which is expressed by only two parameters: p_c and q_c.
- In the GGM, the structure of the communication process is inferred just on the basis of aggregate sales data. The GGM is a parsimonious model.
- The parameters of the GGM are easy to interpret and can provide useful insights from a marketing perspective on the evolution of a given product.

Limitations

- As in the BM and GBM cases, the GGM only deals with the mean behavior of data and is therefore not able to account for other sources of variability, such as seasonality.
- Too short time series make the use of the GGM practically unfeasible or parameter estimates very uncertain. Therefore, the application of the GGM has some limitations in terms of data requirements.

References

Wesley M Cohen and Daniel A Levinthal. Absorptive capacity: A new perspective on learning and innovation. *Administrative Science Quarterly*, 35:128–152, 1990.

Claudia Furlan, Mariangela Guidolin, and Renato Guseo. Has the Fukushima accident influenced short-term consumption in the evolution of nuclear energy? An analysis of the world and seven leading countries. *Technological Forecasting and Social Change*, 107:37–49, 2016.

Mariangela Guidolin and Renato Guseo. Modelling seasonality in innovation diffusion. *Technological Forecasting and Social Change*, 86:33–40, 2014.

Renato Guseo. Strategic interventions and competitive aspects in innovation life cycle. Technical report, Department of Statistical Sciences, University of Padua, 2004.

Renato Guseo and Mariangela Guidolin. Modelling a dynamic market potential: A class of automata networks for diffusion of innovations. *Technological Forecasting and Social Change*, 76(6):806–820, 2009.

Renato Guseo and Mariangela Guidolin. Cellular automata with network incubation in information technology diffusion. *Physica A: Statistical Mechanics and its Applications*, 389(12):2422–2433, 2010.

Renato Guseo and Mariangela Guidolin. Market potential dynamics in innovation diffusion: Modelling the synergy between two driving forces. *Technological Forecasting and Social Change*, 78(1):13–24, 2011.

Vijay Mahajan, Eitan Muller, and Frank M Bass. New product diffusion models in marketing: A review and directions for research. *Journal of Marketing*, 54(1):1–26, 1990.

Geoffrey A Moore. *Crossing the Chasm*. New York: Harper Business, 1999.

Mark Newman. *Networks*. Oxford: Oxford University UPress, 2018.

Carl Shapiro and Hal R Varian. *Information Rules: A Strategic Guide to the Network Economy*. Harvard Business Press, 1999.

5

Dealing with Autocorrelation and Seasonality in Innovation Diffusion

After reading this chapter, you should be able to

- Know the basic features of the ARMAX refinement
- Understand when an ARMAX refinement is needed
- Use the ARMAX refinement for diffusion modeling.

5.1 Introduction

As we have seen so far, innovation diffusion models are concerned with describing the nonlinear mean behavior of a series. Taking the terminology of time series analysis, we could say that innovation diffusion models capture the – nonlinear – *trend* of a series, leaving out other components. This may be enough in strategic terms because these models can predict the time of the maximum peak and the end of the life cycle, they are able to detect the possible effect of external perturbations, and, more importantly, they are interpretable.

However, practical application – especially when dealing with weekly, monthly, or quarterly data – has shown that in many cases the observed series has a very strong variability around the mean, which may be imputable to serial correlation or seasonality. Accounting for variability not adequately captured by the model may be extremely useful, especially for making more accurate short-term predictions.

As an illustrative example, we may reconsider the iPod diffusion, for which the Guseo–Guidolin model (GGM) has been selected as a suitable model, as shown in Figure 5.1. Despite the satisfactory performance of the model in describing the mean behavior of data, it is clear that it completely neglects the strong seasonal component, while accounting for its effect would be in some cases a desirable

Innovation Diffusion Models: Theory and Practice, First Edition. Mariangela Guidolin.
© 2024 John Wiley & Sons Ltd. Published 2024 by John Wiley & Sons Ltd.
Companion Website: www.wiley.com/go/innovationdiffusionmodels/

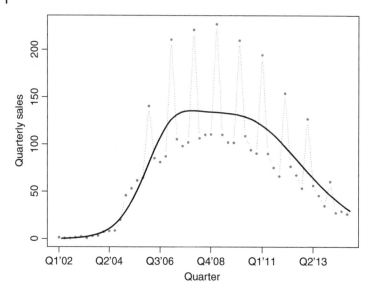

Figure 5.1 GGM for iPod. Quarterly sales.

output of the analysis. To achieve this result, one possible approach, proposed by Guseo (2004) and subsequently applied in many cases in the literature (see, for instance, Guseo et al., 2007, Guseo and Guidolin, 2009, Guseo and Mortarino, 2015, Guidolin et al., 2019), relies on an Autoregressive Moving-Average with external regressor, ARMAX, refinement. The remainder of the chapter is as follows. In Section 5.2, the theory of the ARMAX refinement for innovation diffusion is illustrated, while Section 5.3 shows its direct application to two applied cases, to show how to deal with autocorrelation of residuals and seasonality.

5.2 ARMAX Refinement: Theory

As already seen in Chapters 2–4, a nonlinear regression model may be represented as

$$y(t) = f(\beta, t) + \varepsilon(t),$$

where $y(t)$ is the time series of interest, $f(\beta, t)$ is the deterministic part, and $\varepsilon(t)$ is an error term, assumed to have zero mean, i.e. $E(\varepsilon(t) = 0)$, in order to guarantee the application of the nonlinear least squares (NLS) method. Further assumptions may be made regarding the structure of $\varepsilon(t)$, namely to have constant variance $\text{Var}(\varepsilon(t)) = \sigma^2$ and uncorrelated components, $\sigma_{\varepsilon(t)}, \varepsilon(t') = 0, t \neq t'$. Together these assumptions identify a white noise process, WN $(0, \sigma^2)$.

As a result of the regression modeling, we would expect the residuals to have a behavior coherent with white noise. In particular, we would expect to have no residual autocorrelation. However, the practical application of nonlinear regression models shows that this is often not the case, and instead, residuals typically exhibit positive autocorrelation. This may be detected through some tests, such as the Durbin–Watson test (Durbin and Watson, 1950), or simply by graphically inspecting the behavior of the series of residuals and their autocorrelation function (ACF), (Hyndman and Athanasopoulos, 2018).

In these cases, an improvement of predictions may be performed through an ARMAX refinement. Before entering into the details of this approach, it may be useful to recall the structure of an Autoregressive Moving-Average (h, k), ARMA (h, k), and of ARMAX models. For more details about ARMA (h, k) models see, for instance, Hyndman and Athanasopoulos (2018).

5.2.1 ARMAX Models

For simplicity, we will only focus on nonseasonal ARMA models, but the same concepts may be easily extended to include seasonal terms. Following Hyndman and Athanasopoulos (2018), an ARMA (h, k) model may be defined as

$$y_t = \phi_1 y_{t-1} + \cdots + \phi_h y_{t-h} - \theta_1 z_{t-1} - \cdots - \theta_k z_{t-k} + z_t,$$

where z_t is a white noise process.

An ARMAX model simply adds a term λx_t as follows

$$y_t = \lambda x_t + \phi_1 y_{t-1} + \cdots + \phi_h y_{t-h} - \theta_1 z_{t-1} - \cdots - \theta_k z_{t-k} + z_t, \tag{5.1}$$

where x_t is a covariate at time t and λ is its coefficient. By using the backshift operators, B, Equation (5.1) may be defined in a more compact way as

$$\phi(B)y_t = \lambda x_t + \theta(B)z_t.$$

ARMAX models can be considered as special cases of *transfer function* models (Box et al., 2015).

Following this approach, in Guseo (2004), it is proposed to introduce as a plug-in covariate x_t the predicted values $f(\hat{\beta}, t)$, estimated through NLS procedure, so that Equation (5.1) may be rewritten as

$$y_t = \lambda f(\hat{\beta}, t) + \phi_1 y_{t-1} + \cdots + \phi_h y_{t-h} - \theta_1 z_{t-1} - \cdots - \theta_k z_{t-k} + z_t \tag{5.2}$$

and by rearranging the terms of (5.2)

$$y_t - \lambda f(\hat{\beta}, t) = \phi_1 y_{t-1} + \cdots + \phi_h y_{t-h} - \theta_1 z_{t-1} - \cdots - \theta_k z_{t-k} + z_t.$$

Whenever $\lambda = 1$ we will have

$$y_t - \lambda f(\hat{\beta}, t) = \hat{\varepsilon}(t) = \phi_1 y_{t-1} + \cdots + \phi_h y_{t-h} - \theta_1 z_{t-1} - \cdots - \theta_k z_{t-k} + z_t.$$

Essentially, the inclusion of the covariate $f(\hat{\beta}, t)$ in an ARMAX model is equivalent to applying an ARMA(h, k) model to the series of regression residuals, $\hat{\varepsilon}(t)$. However, the advantage of using the ARMAX model with the estimation of parameter λ is that it includes an evaluation about the suitability of the selected nonlinear model $f(\beta, t)$. If λ is estimated to be ≈ 1, this provides an indication that the nonlinear trend has been efficiently described with $f(\beta, t)$. All these ideas may be easily extended to include seasonality, by using seasonal ARMAX models, SARMAX.

As well known, when using models of the ARMA family, many alternative models can be suitable. The selection may be performed by using standard criteria, such as the Akaike Information Criterion (AIC), (Akaike, 1974).

5.3   ARMAX Refinement: Case Studies

The practical application of the ARMAX (or SARMAX) procedure may be summarized in the following steps:

1. Select a nonlinear model, $f(\beta, t)$, for $y(t)$, e.g. Bass model (BM), Generalized Bass model (GBM), and GGM.
2. Perform an analysis of residuals: graphical inspection of residuals, analysis of ACF. If needed, perform some tests, e.g. Durbin–Watson.
3. If there is autocorrelation in the residuals, apply an ARMAX or SARMAX model directly to $y(t)$, by including $f(\hat{\beta}, t)$ as a plug-in covariate.
4. Check the estimation results, by paying special attention to the estimate of parameter λ.
5. Make a selection of the model based on the AIC.
6. Update the predictions after the refinement.

To show the practical use of this tool, two examples are considered: Netflix subscriptions and sales of Apple iPods.

5.3.1   Netflix Subscriptions

The first example to illustrate the ARMAX refinement procedure is based on the quarterly series of Netflix subscriptions, starting in Q1'13 and ending in Q4'21. The data are available at www.statista.com. Quarterly and corresponding cumulative data are displayed in Figure 5.2, and it is evident that subscriptions have been constantly increasing, so the trend of the series may be reasonably modeled through a simple BM. The results of the BM are graphically reported in Figure 5.3,

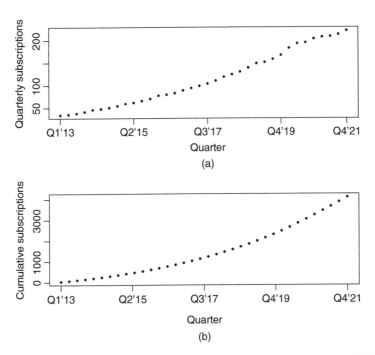

Figure 5.2 Netflix. Quarterly (a) and cumulative (b) subscriptions from 2013 to 2021 (million units).

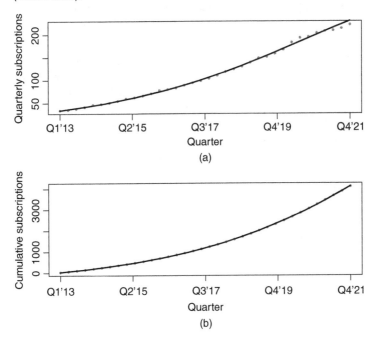

Figure 5.3 BM for Netflix. Quarterly subscriptions (a) and cumulative subscriptions (b).

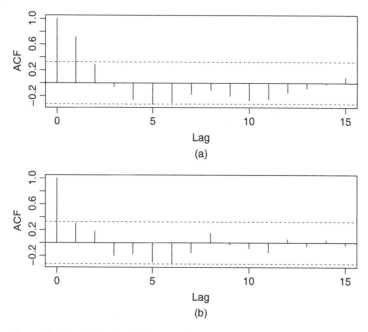

Figure 5.4 Netflix subscriptions: residual ACF after BM (a) and residual ACF after ARMAX refinement (b).

where it may be that a satisfactory fit is obtained. However, one may notice that the quarterly series exhibits a greater variability, from the end of 2019 on, which is less adequately captured by the BM. A confirmation comes from the graphical inspection the ACF, shown on Figure 5.4a. The analysis of the ACF shows that the autocorrelation at lag 1 is significant, and this would justify the need for an ARMAX refinement. To this end, an ARMAX (h, k) model, ARMA(1, 1) with external covariate, $f(\hat{\beta}, t)$, i.e. the fitted values obtained with the BM, is selected. The results are presented in Table 5.1. In particular, the value of parameter λ is practically equal to 1, as desired. The analysis of the residual ACF (after the ARMAX) has no longer significant components, as may be seen in Figure 5.4b. Updating the

Table 5.1 ARMAX refinement for Netflix subscriptions.

Parameter	ar1	ma1	λ
Estimate	0.7665	0.7547	0.9979
s.e.	0.1188	0.0890	0.0019

AIC = 172.34.

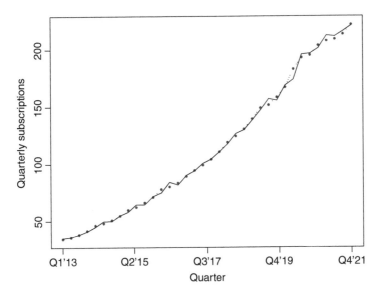

Figure 5.5 ARMAX refinement for Netflix subscriptions.

predictions after the ARMAX refinement produces the clear improvement shown in Figure 5.5.

5.3.2 Apple iPod

A practical application of the SARMAX refinement can be illustrated by considering the case of the Apple iPod. In this case, the presence of a seasonal component is evident, while in other applications the need for a refinement of predictions accounting for seasonality is less obvious. As a first step, we inspect the ACF of the residuals after the GGM, displayed in Figure 5.6a. It is easy to observe a systematic behavior, which is reflected in several significant autocorrelations at the seasonal lags $(4, 8, 12, \ldots)$. The presence of such strong seasonal behavior suggests selecting a SARMAX model. After considering various options, the selected model, according to the AIC, is a SARMAX $(h, k), (H, K)$ model, $(2, 2)(1, 1)$ with external covariate, $f(\hat{\beta}, t)$. Table 5.2 summarizes the estimates of this procedure. Notice in particular the value of parameter $\lambda = 1$, showing that the trend has been estimated without bias. The improvement obtained using the SARMAX model can be based on inspection of the residual ACF, displayed in Figure 5.6b. The update of predictions after the SARMAX is shown in Figure 5.7, where the improvement obtained with respect to the GGM is clearly visible. It should be noticed that the inclusion of the predicted values with the GGM as a covariate has played a crucial role in this refinement procedure.

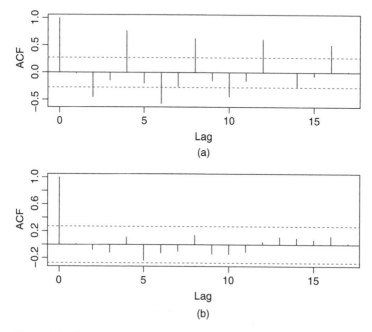

Figure 5.6 iPod. Residual ACF after GGM (a) and residual ACF after SARMAX refinement (b).

Table 5.2 SARMAX refinement for iPod.

Parameter	ar1	ar2	ma1	ma2	sar1	sma1	λ
Estimate	1.5409	−0.8130	−0.9932	−0.0068	0.8586	0.3869	1.0009
s.e.	0.0976	0.0887	0.2033	0.2011	0.8586	0.1512	0.0011

AIC = 404.

5.4 Recap

At the end of this chapter, it is useful to summarize some aspects of the ARMAX refinement.

Properties

- The procedure based on ARMAX or SARMAX refinement allows to account for possible autocorrelation and seasonality in the observed data, which are not captured with an innovation diffusion model, whose primary purpose is to model the mean behavior of the series.

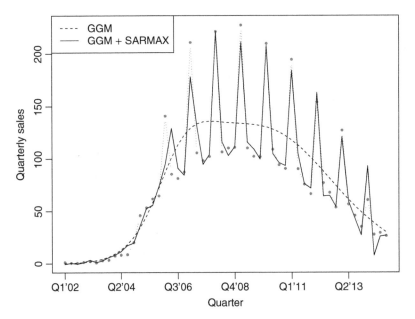

Figure 5.7 SARMAX refinement for iPod and GGM.

- Refining predictions may be useful especially from a short-term perspective when it is crucial to account for seasonal effects or other intra-year movements within observed data.

Limitations

- The ARMAX refinement is a quite simple procedure made of two steps, i.e. the fit of the nonlinear model and the subsequent management of autocorrelation/seasonality. In some cases, this could prove not sufficient for obtaining satisfactory results. A possible evolution of it could envisage the use of iterative procedures that, at each iteration, improve the prediction.

References

Hirotsugu Akaike. A new look at the statistical model identification. *IEEE Transactions on Automatic Control*, 19(6):716–723, 1974.

George EP Box, Gwilym M Jenkins, Gregory C Reinsel, and Greta M Ljung. *Time Series Analysis: Forecasting and Control*. New York: Wiley, 2015.

James Durbin and Geoffrey Watson. Testing for serial correlation in least squares regression: I. *Biometrika*, 37(3/4):409–428, 1950.

Mariangela Guidolin, Renato Guseo, and Cinzia Mortarino. Regular and promotional sales in new product life cycles: Competition and forecasting. *Computers & Industrial Engineering*, 130:250–257, 2019.

Renato Guseo. Strategic interventions and competitive aspects in innovation life cycle. Technical report, Department of Statistical Sciences, University of Padua, 2004.

Renato Guseo and Mariangela Guidolin. Modelling a dynamic market potential: A class of automata networks for diffusion of innovations. *Technological Forecasting and Social Change*, 76(6):806–820, 2009.

Renato Guseo and Cinzia Mortarino. Modeling competition between two pharmaceutical drugs using innovation diffusion models. *The Annals of Applied Statistics*, 9(4):2073–2089, 2015.

Renato Guseo, Alessandra Dalla Valle, and Mariangela Guidolin. World Oil Depletion Models: Price effects compared with strategic or technological interventions. *Technological Forecasting and Social Change*, 74(4):452–469, 2007.

Rob J Hyndman and George Athanasopoulos. *Forecasting: Principles and Practice*. OTexts, 2018.

6

Innovation Diffusion in Competition

After reading this chapter, you should be able to

- Know the basic features of diffusion models in competition
- Interpret the parameters of diffusion models in competition
- Make predictions with diffusion models in competition.

6.1 Introduction

The models presented so far have treated innovation diffusion as a univariate process, whose development depends on its internal parameters and possibly on external perturbations or dynamic market potential, in the form of marketing mix actions, incentive measures, and other structured shocks that are able to alter the speed of adoptions. The univariate approach on which the Bass model (BM), Generalized Bass model (GBM), and Guseo–Guidolin model (GGM) rely is however limited in not explicitly considering the competitive dynamics that characterize almost all commercial and technological markets. In this sense, univariate models may provide a reduced description of the market.

Competition is indeed a crucial factor that may represent both a barrier and an opportunity for a product's life cycle. On the one hand, the competition exerted by a concurrent product is clearly an obstacle to a product's sales growth and may thus cause its failure in the market. On the other hand, competition may also imply some benefits, for example, by increasing the size of a market for a specific commercial category or by reinforcing the knowledge of consumers about it. Hence, accounting for competition has been always considered a crucial element in innovation diffusion modeling (Meade and Islam, 2006, Peres et al., 2010), because it may produce more complete market analyses and a deeper understanding of the forces driving or hindering a diffusion process. Typically,

Innovation Diffusion Models: Theory and Practice, First Edition. Mariangela Guidolin.
© 2024 John Wiley & Sons Ltd. Published 2024 by John Wiley & Sons Ltd.
Companion Website: www.wiley.com/go/innovationdiffusionmodels/

diffusion modeling under competition has been based on the Lotka–Volterra models, a class of models developed by the contributions of Lotka (1920) and Volterra (1926). These equations, originally developed and applied in the natural sciences for describing interactions between species and especially the so-called "predator-prey" relationship, have then also been used in the technological domain. Starting with pioneering works by Abramson and Zanette (1998) and Morris and Pratt (2003), the Lotka–Volterra models have been used to study technological competition and substitution (Guidolin and Guseo, 2020).

In parallel, the marketing literature has expanded to analyze the problem of competition, by extending the structure of the BM. This generalization into a competitive environment, by accounting for more than one diffusion process, has typically focused on bivariate models, where the interaction between two products or technologies is analyzed. The fact of considering no more than two products has to do with the complexity of dealing with systems of differential equations, with an increasing number of parameters to estimate (Guidolin and Guseo, 2020). In the literature, examples of innovation diffusion models in competition, extending the approach based on the BM, have been developed by Krishnan et al. (2000), Savin and Terwiesch (2005), and Guseo and Mortarino (2012, 2014, 2015). For a recent review, see Petropoulos et al. (2022), Section 2.3.20.

A specific and common thread of these models is the introduction of a complex imitation component, which may be divided into two parts: the *within-product* imitation, which is due to a product's specific sales, and the *cross-product* imitation, which accounts for the effect of sales of the competitor on a product's life cycle. This chapter is dedicated to the description of the diffusion model for competition developed by Guseo and Mortarino (2014), the *unbalanced competition and regime change diachronic* (UCRCD) model. This model describes a general competition situation where two products enter the market at different times (diachronic competition). Such a model has been then extended by Guseo and Mortarino (2015) and Guidolin and Guseo (2015), by relaxing some of its critical assumptions, regarding the market potential and the residual market.

The chapter is organized as follows. Section 6.2 describes the main features of the UCRCD model, while two possible extensions of it are illustrated in Section 6.2.1. Two applied case studies are discussed in Section 6.3, highlighting the flexibility of the UCRCD in capturing different competitive environments. Important properties and some limitations are listed in Section 6.4.

6.2 UCRCD Model: Theory

The generalizing model developed by Guseo and Mortarino (2014), UCRCD, considers the competition between two products, that enter the market at different times, the so-called *diachronic competition*. If two products enter the market at the same time, we instead talk about *synchronic competition*, an easier case in modeling terms, but that is less frequent in reality. For this reason, the UCRCD model assumes that the diffusion process is characterized by two stages: an initial one where only one product is in the market, and a second one when the second product enters the market giving rise to competition.

The second product enters the market at time $t = c$, with $c > 0$. The model is structured as a system of two differential equations, with $z_1'(t)$ and $z_2'(t)$ being instantaneous adoptions of the first and second products, respectively

$$
\begin{aligned}
z_1'(t) &= \left\{ \left[p_{1a} + q_{1a}\frac{z(t)}{m} \right] (1 - I_{t>c}) \right. \\
&\quad \left. + \left[p_{1c} + (q_{1c} + \delta)\frac{z_1(t)}{m} + q_{1c}\frac{z_2(t)}{m} \right] I_{t>c} \right\} [m - z(t)], \qquad (6.1) \\
z_2'(t) &= \left[p_2 + (q_2 - \gamma)\frac{z_1(t)}{m} + q_2\frac{z_2(t)}{m} \right] [m - z(t)] I_{t>c}, \\
m &= m_a(1 - I_{t>c}) + m_c I_{t>c}, \\
z(t) &= z_1(t) + z_2(t) I_{t>c}.
\end{aligned}
$$

As a first relevant element in System (6.1), the market potential is different, depending on the phase considered: m_a is the market potential of the first product in the first phase, (i.e. when there is no competition yet), and m_c is the market potential in competition when the second product has entered the market. The residual market, $m - z(t)$, is common to the two products, and $z(t) = z_1(t) + z_2(t)$ represents overall cumulative adoptions. Also, note that $I_{t>c}$ is an indicator function that takes value 1 when $t > c$.

In the first phase, when $t \leq c$, the life cycle of the first product, $z_1'(t)$, is modeled for simplicity according to a BM with the parameters p_{1a}, q_{1a}, and m_a. In competition, when $t > c$, the diffusion of each product is still described according to a BM to which a competition term is added, to account for the influence of the competing product.

In competition, the life cycle of the first product has new parameters: the innovation coefficient under competition, p_{1c}; the *within* imitation coefficient, $q_{1c} + \delta$, which is multiplied by z_1/m and describes imitative behavior due to internal dynamics; and the *cross* imitation coefficient, q_{1c}, multiplied by z_2/m, which provides a measure of the influence of the sales of the second product on the first. The second product is described in a symmetric way with other three parameters: the innovation coefficient, p_2; the *within* imitation coefficient, q_2, describing imitative behavior due to internal sales z_2/m; and the *cross* imitation coefficient, $q_2 - \gamma$, which measures the influence of sales of the first product on the second. Table 6.1 summarizes the parameters involved in the UCRCD model.

Table 6.1 UCRCD model: parameters and description.

Parameter	Description
m_a	Market potential of 1 before competition
p_{1a}	Innovation of 1 before competition
q_{1a}	Imitation of 1 before competition
m_c	Market potential in competition
p_{1c}	Innovation of 1 in competition
$q_{1c} + \delta$	Within imitation of 1 in competition
q_{1c}	Cross imitation of 2 on 1
p_2	Innovation of 2
q_2	Within imitation of 2
$q_2 - \gamma$	Cross imitation of 1 on 2

The relationship between the first and the second product is controlled by the cross imitation coefficients, q_{1c} and $(q_2 - \gamma)$. In principle, one would expect q_{1c} and $(q_2 - \gamma)$ to be negative, coherently with a negative effect implied by competition. However, practical experience with the application of the UCRCD has shown that parameters may be also estimated to be positive so that the presence of a competitor would be beneficial in some cases. This flexibility gives rise to a richer set of possible relationships between the two products: negative cross imitation coefficients imply competitive dynamics, whereas positive cross imitation coefficients express a sort of "mutualistic" relationship, for simplicity termed "collaboration." Mixed cases are also possible. Table 6.2 illustrates such relationships (Guidolin and Alpcan, 2019, Bessi et al., 2021). A noteworthy aspect refers to the parameters δ and γ, according to what is proposed by Guseo and Mortarino (2014): if these are assumed to be equal, i.e. $\delta = \gamma$, the model takes a reduced form, termed *standard* UCRCD (Guseo and Mortarino, 2014). This assumption implies

Table 6.2 Sign of cross imitation coefficient and their meaning.

q_{1c}	$(q_2 - \gamma)$	Type of relationship
−	−	Full competition
−	+	2 competes with 1, 1 collaborates with 1
+	−	2 collaborates with 1, 1 competes with 2
+	+	Full collaboration

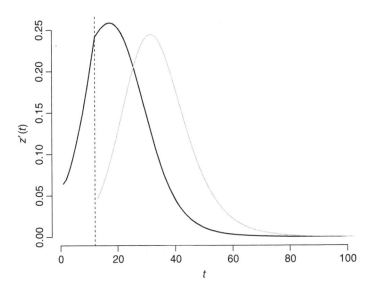

Figure 6.1 UCRCD model: $m_a = 4$, $p_{1a} = 0.008$, $q_{1a} = 0.161$, $m_c = 12$, $p_{1c} = 0.020$, $p_2 = 0.003$, $q_{1c} = -0.060$, $q_2 = 0.205$, $\delta = 0.12$, $\gamma = 0.208$. The second product enters at time $c = 12$ (dashed line).

a symmetry between the two competitors. One of the interesting consequence of this restriction is the possibility to have a closed-form solution for the model (for the details, see Guseo and Mortarino, 2014). Instead, if δ and γ are not constrained, the UCRCD takes a more flexible form, called *unrestricted* UCRCD, for which no closed-form solution is available. By not imposing constraints, the unrestricted UCRCD model in some cases may imply a better and more realistic description of competitive dynamics.

Figure 6.1 describes the trajectories of two products in competition, according to an unrestricted UCRCD model ($\delta \neq \gamma$). The black line describes the behavior of the first product, $z_1'(t)$, which presents a discontinuity due to the presence of two sub-models: a simple BM in the first phase, and the UCRCD model in the second. The second product, $z_2'(t)$ is described by the light gray line. The vertical dashed line displays the time of the entrance of the second product, $c = 12$. Concerning the parameters, these have been set as follows: $m_a = 4$, $p_{1a} = 0.008$, $q_{1a} = 0.161$, $m_c = 12$, $p_{1c} = 0.020$, $p_2 = 0.003$, $q_{1c} = -0.060$, $q_2 = 0.205$, $\delta = 0.12$, $\gamma = 0.208$.

6.2.1 More General Models in Competition

This section briefly outlines two models for competition that generalize the structure of the UCRCD and have been developed by modifying the structure of the residual market, and of the market potential, respectively. It is also worth recalling a further extension of the UCRCD model introduced by Furlan and Mortarino (2018), who proposed a competition model, which accounts for the possible presence of external perturbations in the form of structured shocks $x(t)$, by taking inspiration from the GBM for the univariate case.

6.2.1.1 Lotka–Volterra with Churn Model, LVch

The model presented in this section takes advantage of the structure of the Lotka–Volterra equations to build a more flexible model which, under special conditions, reduces to the UCRCD model. One of the characterizing elements of the UCRCD is that the two products have a common residual market, given by the sum of each product's residual market. This assumption may be relaxed by setting that each product has its own residual market, plus a fraction of the other's, based on the hypothesis that one product is able to "steal" a portion of the residual market of the other, giving rise to a sort of "churn" effect. This idea has been introduced by Guidolin and Guseo (2015), with the Lotka–Volterra with churn model, (LVch).

Similarly to the UCRCD model, the LVch is based on a system of differential equations that describe two life cycles, through a diachronic competition

$$z_1'(t) = \left[p_{1a} + q_{1a} \frac{z_1(t)}{m_a} \right] \left[m_a - z_1(t) \right], \quad 0 \leqslant t \leqslant c,$$

$$z_1'(t) = \left[p_1 + \frac{a_1 z_1(t) + \gamma_2 b_1 z_2(t)}{m_1 + \gamma_2 m_2} \right] \left[(m_1 - z_1(t)) + \gamma_2 (m_2 - z_2(t)) \right], \qquad (6.2)$$

$$z_2'(t) = \left[p_2 + \frac{a_2 z_2(t) + \gamma_1 b_2 z_1(t)}{m_2 + \gamma_1 m_1} \right] \left[(m_2 - z_2(t)) + \gamma_1 (m_1 - z_1(t)) \right].$$

As in the UCRCD model, the first equation of Equation (6.2) describes the life cycle of the first product when $t \leq c$, (i.e. when the competition has not started yet).

The second and third equations describe competition, for $t > c$, when the second product has entered the market. As mentioned, in this model, the residual market is the sum of the product-specific one, $m_i - z_i(t)$, plus a fraction of the other's, $\gamma_j[m_j - z_j(t)]$. The parameters $\gamma_j, j = 1, 2$, modulate the size of this fraction.

As already seen in the UCRCD, $p_i, i = 1, 2$, describe the innovative behavior, whereas the imitation is made of within-product component $[a_1 z_1(t)/(m_1 + \gamma_2 m_2)]$ and a cross-product one, $[\gamma_2 b_1 z_2(t)/(m_1 + \gamma_2 m_2)]$, for the first competitor and, symmetrically, $[a_2 z_2(t)/(m_2 + \gamma_1 m_1)]$ and $[\gamma_1 b_2 z_1(t)/(m_2 + \gamma_1 m_1)]$ for the second. Depending on the value taken by γ_1 and γ_2, the LVch may be reduced to some noteworthy situations:

1. if $0 < \gamma_1 < 1$ and $0 < \gamma_2 < 1$, we have the full LVch model, where both products are affected by within-product and cross-product imitation, and each one may have access to a portion of the other's residual market;

Table 6.3 LVch model reductions depending on values of γ_1 and γ_2.

	Product 1	Product 2
LVch	$0 < \gamma_2 < 1$	$0 < \gamma_1 < 1$
UCRCD	$\gamma_2 = 1$	$\gamma_1 = 1$
Independent BM	$\gamma_2 = 0$	$\gamma_1 = 0$
Direct cannibalization	$\gamma_2 = 0$	$\gamma_1 = 1$
Inverse cannibalization	$\gamma_2 = 1$	$\gamma_1 = 0$

2. if $\gamma_1 = \gamma_2 = 1$, the LVch model reduces to the UCRCD. In this case, the market potential is a common element $m = m_1 + m_2$, and the residual $m - z(t)$, with $z(t) = z_1(t) + z_2(t)$, is completely accessible to both competitors,
3. if $\gamma_1 = \gamma_2 = 0$, no competition exists between the two products, whose life cycle is given by two independent BMs.

Table 6.3 summarizes these possibilities based on the values taken by the parameters γ_1 and γ_2. Mixed cases, i.e. $\gamma_1 = 0, \gamma_2 = 1$ and $\gamma_1 = 1, \gamma_2 = 0$ have been also considered by Guidolin and Guseo (2020), identifying phenomena of cannibalization (direct and inverse cannibalization). A particular application of the LVch model has been developed by Guidolin et al. (2019) to study the possible competition between regular and promotional sales of the same product.

6.2.1.2 Competition Dynamic Market Potential

Another more general model for competition has been developed by Guseo and Mortarino (2015), to account for the presence of a dynamic market potential. Unlike the UCRCD, this model considers a synchronic competition and presents the following structure

$$z_1'(t) = m(t) \left[p_{1c} + (q_1 + \delta)\frac{z_1(t)}{m(t)} + q_1\frac{z_2(t)}{m(t)} \right] \left[1 - \frac{z(t)}{m(t)} \right] + z_1(t)\frac{m'(t)}{m(t)}, \quad (6.3)$$

$$z_2'(t) = m(t) \left[p_2 + (q_2 - \gamma)\frac{z_1(t)}{m(t)} + q_2\frac{z_2(t)}{m(t)} \right] \left[1 - \frac{z(t)}{m(t)} \right] + z_2(t)\frac{m'(t)}{m(t)}.$$

As may be easily observed, the structure of the model (6.3) is similar to that of the UCRCD for the competition part. The form proposed for $m(t)$ in the applied case studied by Guseo and Mortarino (2015) is the one introduced by Guseo and Guidolin (2009) in the GGM, but other structures for $m(t)$ are also considered in their article.

6.2.1.3 Competition Between Three Products, UCTT

So far, we have seen models that try to describe the competitive dynamics between two products or technologies. The possibility to model more than two series has received a limited attention, because of the difficulty of treating systems with three or more differential equations. However, the interaction between three products has been recently considered in some research outputs and some models have been developed to try to account for the presence of three competitors. This section presents the structure of a model for three competitors proposed in Savio et al. (2022), called Unbalanced Competition for Three Technologies, UCTT. The model structure has a common thread with the 3PM model proposed by Furlan et al. (2021) but also has some crucial differences, which will be highlighted at the end of the section.

The UCTT is characterized by two phases: the first one, called *competition phase*, where two products enter the market at the same time and interact with each other, and the second one, called *double-competition phase*, where the third competitor enters the market and interacts with the previous two products.

The market potential $m = m_c + m_d$, i.e. the maximum level of consumption for the three products, has two different levels: m_c, the market potential in the *simple-competition phase*, when the first two products are in the market, and m_d, the global market potential in the *double-competition phase*, when the third product has entered the market. The model takes a common residual market $m - z(t)$ for each competitor, with $z(t) = z_1(t) + z_2(t) + z_3(t)$ the common cumulative consumption, and $z_i(t)$, $i = 1, 2, 3$ the cumulative consumption of product i. The model is a system of differential equations where $z'_1(t)$, $z'_2(t)$, and $z'_3(t)$ represent instantaneous consumption of the first, the second, and the third product, respectively. The third competitor enters the market in $t = a$, with $a > 0$, creating two different phases, i.e. simple competition and double competition.

$$
\begin{aligned}
z'_1(t) = \Bigg\{ &\left[p_{1c} + (q_{1c} + \delta)\frac{z_1(t)}{m} + q_{1c}\frac{z_2(t)}{m} \right] I_{t<a} \\
&+ \left[p_{1d} + (q_{1d} + \varsigma)\frac{z_1(t)}{m} + q_{1d}\frac{z_2(t) + z_3(t)}{m} \right] I_{t \geq a} \Bigg\} [m - z(t)],
\end{aligned}
$$

$$
\begin{aligned}
z'_2(t) = \Bigg\{ &\left[p_{2c} + (q_{2c} - \gamma)\frac{z_1(t)}{m} + q_{2c}\frac{z_2(t)}{m} \right] I_{t<a} \\
&+ \left[p_{2d} + q_{2d}\frac{z_2(t)}{m} + (q_{2d} - \rho)\frac{z_1(t) + z_3(t)}{m} \right] I_{t \geq a} \Bigg\} [m - z(t)],
\end{aligned}
\qquad (6.4)
$$

$$
z'_3(t) = \Bigg\{ \left[p_{3d} + q_{3d}\frac{z_3(t)}{m} + (q_{3d} - \xi)\frac{z_1(t) + z_2(t)}{m} \right] I_{t \geq a} \Bigg\} [m - z(t)],
$$

$$m = m_c I_{t<a} + m_d I_{t \geq a},$$
$$z(t) = z_1(t) + z_2(t) + z_3(t) I_{t \geq a}.$$

Parameters referred to as the *simple-competition phase* are indicated with the subscript c, while those referred to as the *double-competition phase* are indicated with the subscript d.

The system describes complex dynamics between the three products that could have both a competitive and/or collaborative nature. Each phase is characterized by new parameters referred to as both the internal and crossed dynamics for each product. The internal growth is described through the innovation coefficient, which represents the initial spread of the product due to innovative consumers, and the within imitation coefficient, which describes the internal growth of the product due to word-of-mouth. Interaction between competitors is described through the cross imitation coefficient, which measures the effect of the competitors on the growth of the considered product in terms of positive or negative influence. Parameters of the simple-competition phase are illustrated in Table 6.4, while those of the double-competition phase are contained in Table 6.5. Under the double-competition phase, the coefficients describe the joint effect of two products on the third. It is important to observe that the model assumes that two products have the same influence on the third: this is indeed a simplifying assumption but allows us to manage the system of differential equations, which otherwise would become over-parameterized.

Since the UCTT has been inspired by the 3PM model proposed by Furlan et al. (2021) it is worth observing the differences between the two. The 3PM model has

Table 6.4 UCTT model (simple competition): parameters and description.

Parameter	Description
m_c	Market potential in simple competition
p_{1c}	Innovation of 1
$q_{1c} + \delta$	Within imitation of 1
q_{1c}	Cross imitation of 2 on 1
p_{2c}	Innovation of 2
q_{2c}	Within imitation of 2
$q_{2c} - \gamma$	Cross imitation of 1 on 2

Table 6.5 UCTT model (double competition): parameters and description.

Parameter	Description
m_d	Market potential in double competition
p_{1d}	Innovation of 1
$q_{1d} + \zeta$	Within imitation of 1
q_{1d}	Cross imitation of 2, 3 on 1
p_{2d}	Innovation of 2
q_{2d}	Within imitation of 2
$q_{2d} - \rho$	Cross imitation of 1, 3 on 2
p_{3d}	Innovation of 3
q_{3d}	Within imitation of 3
$q_{3d} - \xi$	Cross imitation of 1, 2 on 3

the following structure

$$z_1'(t) = m \left\{ \left[p_{1\alpha} + (q_{1\alpha} + \delta_\alpha) \frac{z_1(t)}{m} + q_{1\alpha} \frac{z_2(t)}{m} \right] (1 - I_{t>c_2}) \right.$$

$$\left. + \left[p_{1\beta} + (q_{1\beta} + \delta_\beta) \frac{z_1(t)}{m} + q_{1\beta} \frac{z_2(t) + z_3(t)}{m} \right] I_{t>c_2} \right\} \left[1 - \frac{z(t)}{m} \right],$$

$$z_2'(t) = m \left\{ \left[p_{2\alpha} + (q_{2\alpha} - \delta_\alpha) \frac{z_1(t)}{m} + q_{2\alpha} \frac{z_2(t)}{m} \right] (1 - I_{t>c_2}) \right.$$

$$\left. + \left[p_{2\beta} + (q_{2\beta} + \delta_\beta) \frac{z_2(t)}{m} + q_{2\beta} \frac{z_1(t) + z_3(t)}{m} \right] I_{t>c_2} \right\} \left[1 - \frac{z(t)}{m} \right],$$

$$z_3'(t) = m \left\{ \left[p_3 + (q_3 - \delta_\beta) \frac{z_1(t) + z_2(t)}{m} + q_3 \frac{z_3(t)}{m} \right] I_{t>c_2} \right\} \left[1 - \frac{z(t)}{m} \right],$$

$$m = m_\alpha (1 - I_{t>c_2}) + m_\beta I_{t>c_2},$$

$$z(t) = z_1(t) + z_2(t) + z_3(t) I_{t>c_2}.$$

The UCTT may be considered as a generalization of the 3PM model, since it uses ζ, ρ, and ξ to distinguish the effects of the cross imitation effect on each of the three products, whereas the 3PM model makes a more restricted assumption. In particular, if the restriction $\zeta = \rho = \xi$ applies on the UCCT model, this leads to the 3PM model.

6.3   UCRCD Model: Case Studies

The UCRCD model has been employed to study competition in different industrial sectors, such as the pharmaceutical market (Guseo and Mortarino, 2014) and the music industry (Guidolin and Guseo, 2015). Several applications of the UCRCD have been made in the energy sector, for example, in Guidolin and Guseo (2016), Furlan and Mortarino (2018), Guidolin and Alpcan (2019), and Bessi et al. (2021), to study competition dynamics between energy sources.

In this section, we propose two applications of the UCRCD model to the energy transition occurring in two countries, Denmark and Australia. Specifically, the relationship between renewables and gas is analyzed in the case of Denmark, while the interplay between coal and renewables is considered for Australia. The case of Denmark has been analyzed by Bessi et al. (2021) and is here updated by considering the 2020 data point. The case of Australia has been studied in Guidolin and Alpcan (2019) and is here updated with four new data points. The general idea of these two applications is to study and estimate the possible competitive strength of renewables toward a fossil fuel, mostly used for electricity production (gas in Denmark and coal in Australia). The data are publicly available at www.bp.com and for all the series refer to annual consumption of the energy source, measured in Exajoule.

6.3.1   Model Fit

In terms of model estimation, the procedure adopted is different with respect to what is seen in the univariate cases of BM, GBM, and GGM, where the available closed-form solution has been applied to cumulative data. Since the unrestricted UCRCD model does not have a closed-form solution, similarly to most bivariate diffusion models of the Lotka–Volterra family, the UCRCD is estimated by directly using the differential equations of Equation (6.1) to instantaneous data (in this case annual data). The applied model is therefore

$$w_1'(t) = m_a \left[p_{1a} + q_{1a} \frac{z(t)}{m_a} \right] + \varepsilon_1(t), \quad t < c,$$

$$w_1'(t) = m_c \left[p_{1c} + (q_{1c} + \delta) \frac{z_1(t)}{m_c} + q_{1c} \frac{z_2(t)}{m_c} \right] \left[1 - \frac{z(t)}{m_c} \right] + \varepsilon_1(t),$$

$$w_2'(t) = m_c \left[p_2 + (q_2 - \gamma) \frac{z_1(t)}{m_c} + q_2 \frac{z_2(t)}{m_c} \right] \left[1 - \frac{z(t)}{m_c} \right] + \varepsilon_2(t),$$

where $\varepsilon_1(t)$ and $\varepsilon_2(t)$ are error terms.

As in the univariate case, estimates are performed via nonlinear least squares (NLS), for which some starting values of the involved parameters are required. This is typically not a trivial procedure, because for most parameters there is no clear indication on how to set them. However, for parameters $m_a, p_{1a},$ and $q_{1a},$ it is reasonable to use the estimates of a BM applied to the first product, while the competition phase requires more careful management and some working hypotheses on the magnitude of the parameters.

6.3.2 Denmark

For the case of Denmark, the series of gas goes from 1985 to 2020, while renewables are considered from 1997 to 2020. Figure 6.2a well suggests the possible competition between renewables and gas, and the progressive decline of gas may be related to the parallel growth of renewables. This is especially visible starting from the 2010s. In order to study these dynamics, an unrestricted UCRCD ($\delta \neq \gamma$) has been estimated, the results of which are displayed in Table 6.6 and Figure 6.3.

A first inspection of the estimates confirms a satisfactory model selection. Parameters $m_a, p_{1a},$ and q_{1a} are referred to the BM used to describe the first phase of gas consumption, until the year 1996. As may be observed, the entrance of renewables in 1997 has increased the market potential which, in competition, is $m_c = 13.46$. Thus, in terms of the size of the market, the competition seems to have exerted a positive effect. Obviously, the most interesting results refer to the imitation components, within ($q_{1c} + \delta, q_2$) and cross ($q_{1c}, q_2 - \gamma$). As concerns within imitation, the values $q_{1c} + \delta = 0.057$ and $q_2 = 0.18$ confirm a positive dynamics of growth, indeed much stronger in the case of renewables. Instead, the cross imitation parameters, $q_{1c} = -0.074$ and $q_2 - \gamma = -0.02$ are both negative, confirming the existence of a competitive relationship, as initially hypothesized.

Table 6.6 Parameter estimates of UCRCD for gas renewables in Denmark.

Parameter	Estimate	s.e.	Lower c.i.	Upper c.i.
m_a	4.33	2.018	0.37	8.28
p_{1a}	0.008	0.003	0.001	0.014
q_{1a}	0.16	0.020	0.12	0.20
m_c	13.46	2.33	8.88	18.03
p_{1c}	0.009	0.0018	0.006	0.013
q_{1c}	−0.074	0.0159	−0.105	−0.042
q_2	0.18	0.063	0.061	0.31
δ	0.13	0.021	0.089	0.172
γ	0.20	0.085	0.036	0.37

$R^2 = 0.9752.$

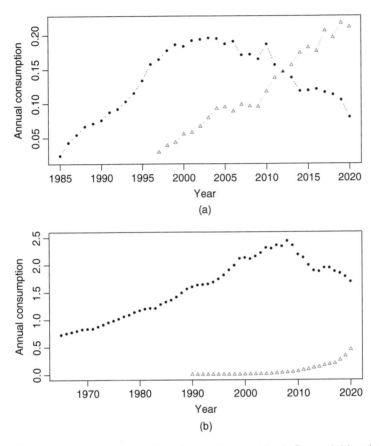

Figure 6.2 Annual consumption of gas and renewables in Denmark (a) and annual consumption of coal and renewables in Australia (b).

Also, the competitive strength of renewables on gas, q_{1c}, is higher than the one of gas on renewables, $q_2 - \gamma$.

6.3.3 Australia

For the case of Australia, the time series of coal consumption has been considered from 1965 to 2020, while the first observation available for renewables is in 1990. A first look at the data in part (b) of Figure 6.2 suggests a very different situation with respect to Denmark. Renewables have started to grow in the last decade (from 2010) and coal seems to have a dominant role in the market, although its consumption has been markedly declining since 2007. In this case, the visual inspection of

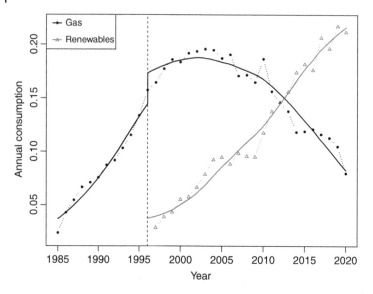

Figure 6.3 UCRCD model for gas and renewables consumption in Denmark. The dotted vertical line indicates when the competitor entered the market.

data is not particularly suggestive of the existence of a relationship, and the application of the UCRCD model may be useful for capturing the dynamics between the two energy sources if any. The resulting best model is the standard UCRCD ($\delta = \gamma$), as reported in Table 6.7 and Figure 6.4. Focusing on the parameters, one may notice that those of the first phase are poorly estimated and in particular the market potential m_a is not significant. This is essentially due to the fact that the series of coal is not available since its beginning and this may cause some issues in the BM estimation. However, the most interesting results come from the analysis of the competition phase. The first element of interest is that the innovation coefficient for renewables is not significant and for this reason not reported in the table. Such finding is coherent with similar results reported in the literature about the fragile role of innovators in the diffusion of renewable energy sources (Guidolin and Mortarino, 2010, Guidolin and Alpcan, 2019, Bunea et al., 2020, 2022, Bessi et al., 2021).

Regarding imitation, the within imitation effect is positive for both energy sources, $q_{1c} + \delta = 0.07$ and $q_2 = 0.529$, with a more evident growth of renewables. The analysis of cross imitation parameters reflects a mixed situation, i.e. $q_{1c} = -0.455$ and $q_2 - \delta = 0.0004$: renewables strongly compete with coal, while coal has a positive, although extremely weak, effect on renewables.

Table 6.7 Parameter estimates of UCRCD for coal renewables in Australia.

Parameter	Estimate	s.e.	Lower c.i.	Upper c.i.
m_a	491.37	273.86	−45.38	1028.10
p_{1a}	0.0014	0.001	−0.000	0.003
q_{1a}	0.034	0.0017	0.030	0.037
m_c	128.64	3.40	121.97	135.32
q_{1c}	−0.455	0.067	−0.587	−0.322
q_2	0.529	0.065	0.400	0.657
δ	0.528	0.066	0.397	0.659

$R^2 = 0.9955$.

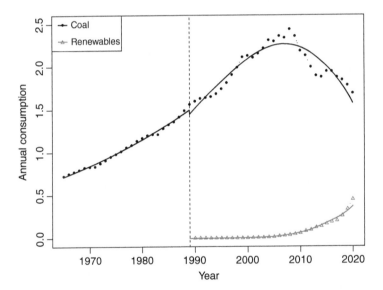

Figure 6.4 UCRCD model for coal and renewables consumption in Australia. The dotted vertical line indicates when the competitor entered the market.

6.4 Recap

At the end of this chapter, some interesting properties of the UCRCD may be recalled.

Properties

- The UCRCD model allows to account for the interplay between two products or technologies through a suitable description of the imitation, which is split within and a cross components
- The UCRCD model is able to describe competitive (or collaborative) dynamics through a limited number of parameters: it is a parsimonious model
- The UCRCD model is easy to use and easy to interpret, because all the parameters have a clear interpretation.

Limitations

The UCRCD model (along with its extensions) is limited in only describing competition between pairs of products. Some proposals to account for more than two competitors have been made by Furlan et al. (2021) and Savio et al. (2022). However, these models present a high number of parameters, which typically pose serious challenges in terms of identifiability and significance. In order to reduce the number of parameters, one can set some constraints, but this introduces some hypotheses about the competition between products that could be very strong.

References

Guillermo Abramson and Damián H Zanette. Statistics of extinction and survival in Lotka-Volterra systems. *Physical Review E*, 57(4):4572, 1998.

Alessandro Bessi, Mariangela Guidolin, and Piero Manfredi. The role of gas on future perspectives of renewable energy diffusion: Bridging technology or lock-in? *Renewable and Sustainable Energy Reviews*, 152:111673, 2021.

Anita M Bunea, Pompeo Della Posta, Mariangela Guidolin, and Piero Manfredi. What do adoption patterns of solar panels observed so far tell about governments' incentive? Insights from diffusion models. *Technological Forecasting and Social Change*, 160:120240, 2020.

Anita M Bunea, Mariangela Guidolin, Piero Manfredi, and Pompeo Della Posta. Diffusion of solar PV energy in the UK: A comparison of sectoral patterns. *Forecasting*, 4(2):456–476, 2022.

Claudia Furlan and Cinzia Mortarino. Forecasting the impact of renewable energies in competition with non-renewable sources. *Renewable and Sustainable Energy Reviews*, 81:1879–1886, 2018.

Claudia Furlan, Cinzia Mortarino, and Mohammad Salim Zahangir. Interaction among three substitute products: An extended innovation diffusion model. *Statistical Methods & Applications*, 30(1):269–293, 2021.

Mariangela Guidolin and Tansu Alpcan. Transition to sustainable energy generation in Australia: Interplay between coal, gas and renewables. *Renewable Energy*, 139:359–367, 2019.

Mariangela Guidolin and Renato Guseo. Technological change in the US music industry: Within-product, cross-product and churn effects between competing blockbusters. *Technological Forecasting and Social Change*, 99:35–46, 2015.

Mariangela Guidolin and Renato Guseo. The German energy transition: Modeling competition and substitution between nuclear power and renewable energy technologies. *Renewable and Sustainable Energy Reviews*, 60:1498–1504, 2016.

Mariangela Guidolin and Renato Guseo. Has the iPhone cannibalized the iPad? An asymmetric competition model. *Applied Stochastic Models in Business and Industry*, 36(3):465–476, 2020.

Mariangela Guidolin and Cinzia Mortarino. Cross-country diffusion of photovoltaic systems: Modelling choices and forecasts for national adoption patterns. *Technological Forecasting and Social Change*, 77(2):279–296, 2010.

Mariangela Guidolin, Renato Guseo, and Cinzia Mortarino. Regular and promotional sales in new product life cycles: Competition and forecasting. *Computers & Industrial Engineering*, 130:250–257, 2019.

Renato Guseo and Mariangela Guidolin. Modelling a dynamic market potential: A class of automata networks for diffusion of innovations. *Technological Forecasting and Social Change*, 76(6):806–820, 2009.

Renato Guseo and Cinzia Mortarino. Sequential market entries and competition modelling in multi-innovation diffusions. *European Journal of Operational Research*, 216(3):658–667, 2012.

Renato Guseo and Cinzia Mortarino. Within-brand and cross-brand word-of-mouth for sequential multi-innovation diffusions. *IMA Journal of Management Mathematics*, 25(3):287–311, 2014.

Renato Guseo and Cinzia Mortarino. Modeling competition between two pharmaceutical drugs using innovation diffusion models. *The Annals of Applied Statistics*, 9(4):2073–2089, 2015.

Trichy V Krishnan, Frank M Bass, and V Kumar. Impact of a late entrant on the diffusion of a new product/service. *Journal of Marketing Research*, 37(2):269–278, 2000.

Alfred J Lotka. Analytical note on certain rhythmic relations in organic systems. *Proceedings of the National Academy of Sciences of the United States of America*, 6(7):410–415, 1920.

Nigel Meade and Towhidul Islam. Modelling and forecasting the diffusion of innovation–a 25-year review. *International Journal of Forecasting*, 22(3):519–545, 2006.

Steven A Morris and David Pratt. Analysis of the Lotka-Volterra competition equations as a technological substitution model. *Technological Forecasting and Social Change*, 70(2):103–133, 2003.

Renana Peres, Eitan Muller, and Vijay Mahajan. Innovation diffusion and new product growth models: A critical review and research directions. *International Journal of Research in Marketing*, 27(2):91–106, 2010.

Fotios Petropoulos, Daniele Apiletti, Vassilios Assimakopoulos, et al. Forecasting: Theory and practice. *International Journal of Forecasting*, 38(3):705–871, 2022.

Sergei Savin and Christian Terwiesch. Optimal product launch times in a duopoly: Balancing life-cycle revenues with product cost. *Operations Research*, 53(1):26–47, 2005.

Andrea Savio, Luigi De Giovanni, and Mariangela Guidolin. Modelling energy transition in Germany: An analysis through ordinary differential equations and system dynamics. *Forecasting*, 4(2):438–455, 2022.

Vito Volterra. Fluctuations in the abundance of a species considered mathematically. *Nature*, 118(2972):558–560, 1926.

7

Estimation Methods for Innovation Diffusion Models

7.1 Introduction

This chapter is a more technical one, which deals with estimation and inferential aspects for innovation diffusion models.

Many statistical methods have been proposed in literature to estimate parameters of a nonlinear model, such as the ones described in this book. We refer to Chapter 2 of Seber and Wild (1989) for a review of the most used methods, such as maximum-likelihood, quasi-likelihood, robust and Bayesian estimators. Here, we describe the ideas and properties of the nonlinear least squares, which is probably the most used method for innovation diffusion models.

Section 7.2 focuses on the Nonlinear Least Squares approach, and illustrates the Gauss–Newton and the Levenberg–Marquardt algorithms. A general overview on confidence intervals and hypothesis testing is provided in Section 7.3.

7.2 Nonlinear Least Squares

The nonlinear least squares (NLS) is a typical method for nonlinear regression problems. Let us consider a model with fixed regressors in which the functional form f is known and $x_i, y_i, i = 1, 2, \ldots, n$ are the n available couples of observations,

$$y_i = f(x_i; \vartheta^*) + \varepsilon_i, \quad i = 1, 2, \ldots, n. \tag{7.1}$$

In Equation (7.1), we assume $E(\varepsilon_i) = 0$ and define $x_i \in \mathbb{R}^k$ a k-dimensional vector of explanatory variables, $y_i \in \mathbb{R}$ the response variable, ϑ^* the unknown true parameter ϑ, where $\vartheta \in \Theta \subset \mathbb{R}^p$. A nonlinear least squares estimate $\hat{\vartheta}$ of the unknown parameter ϑ^* is obtained by minimizing, for $\vartheta \in \Theta$, the deviance $S(\vartheta)$

$$S(\vartheta) = \sum_{i=1}^{n} [y_i - f(x_i; \vartheta)]^2. \tag{7.2}$$

Innovation Diffusion Models: Theory and Practice, First Edition. Mariangela Guidolin.
© 2024 John Wiley & Sons Ltd. Published 2024 by John Wiley & Sons Ltd.
Companion Website: www.wiley.com/go/innovationdiffusionmodels/

Unlike the linear case, we may often obtain for $S(\vartheta)$ different local minima. Also, although NLS estimate $\hat{\vartheta}$ is a biased estimate, if ε_i are independent and identically distributed with variance $V(\varepsilon) = \sigma^2$, $i = 1, 2, \ldots, n$, under suitable regularity conditions $\hat{\vartheta}$ and $S(\hat{\vartheta})/(n - p)$, are consistent estimates of ϑ^* and σ^2, respectively. Additional regularity conditions ensure that $\hat{\vartheta}$ is asymptotically multinormal. By assuming normality of ε_i, $i = 1, 2, \ldots, n$, $\hat{\vartheta}$ is also the maximum likelihood estimate.

The optimization of $S(\vartheta)$, $\vartheta \in \Theta$, does not have a closed form and requires iterative procedures, which pose computational problems. However, if $f(x_i; \vartheta)$ is differentiable at ϑ and $\hat{\vartheta}$ is an internal point of Θ, then the problem is solved by differentiating and posing equal to zero $S(\vartheta)$.

By denoting $f(\vartheta) = [f(x_1, \vartheta), f(x_2, \vartheta), \ldots, f(x_n, \vartheta)]^\top$ the vector of functions at observed points $x_1, x_2, \ldots, x_n$, and $y = [y_1, y_2, \ldots, y_n]$ the vector of responses, respectively, $S(\vartheta)$ may be rewritten

$$S(\vartheta) = [y - f(\vartheta)]^\top [y - f(\vartheta)].$$

Coherently, the normal equations $f(\vartheta)$ are defined as

$$-2J^\top [y - f(\vartheta)] = 0,$$

where $J = \partial f(\vartheta)/\partial \vartheta$ is the Jacobian of $f(\vartheta)$.

7.2.1 Gauss–Newton Method

Consider an initial vector of parameters ϑ_a that may be a good approximation of $\hat{\vartheta}$ and the first-order Taylor approximation of the response vector $f(\vartheta)$ in a neighborhood of the point ϑ_a

$$f(\vartheta) \approx f(\vartheta_a) + J_a(\vartheta - \vartheta_a), \tag{7.3}$$

where J_a is the Jacobian at ϑ_a.

Based on Equation (7.3), the deviance may be approximated linearly

$$\begin{aligned}
S(\vartheta) &= [y - f(\vartheta)]^\top [y - f(\vartheta)] \\
&\approx [y - f(\vartheta_a) - J_a(\vartheta - \vartheta_a)]^\top [y - f(\vartheta_a) - J_a(\vartheta - \vartheta_a)] \\
&= [z - J_a \beta]^\top [z - J_a \beta], \tag{7.4}
\end{aligned}$$

where $z = y - f(\vartheta_a)$ and $\beta = (\vartheta - \vartheta_a)$. The minimum $\hat{\beta}$ is easy to obtain because Equation (7.4) is locally linear, $\hat{\beta} = (J_a^\top J_a)^{-1} J_a^\top z$, and, therefore,

$$\hat{\beta} = (\vartheta_b - \vartheta_a) = (J_a^\top J_a)^{-1} J_a^\top [y - f(\vartheta_a)] = \delta_a.$$

The second approximation of $\hat{\vartheta}$ is thus a correction of the starting point ϑ_a through δ_a,

$$\vartheta_b = \vartheta_a + \delta_a,$$

and this process may be iterated until convergence. It can be shown that in correspondence to a solution, the iterative correction δ_a decreases to zero and stops the updating procedure. The major strengths of Gauss–Newton-based algorithms come from their requiring only the first derivatives.

7.2.2 Levenberg–Marquardt Method

The method of Levenberg–Marquardt introduces a modification in the algorithm of Gauss–Newton by eliminating sources of singularity due to the matrix $(J_a^\top J_a)$ through a suitable diagonal matrix of full rank D_a with positive elements

$$\delta_a = (J_a^\top J_a + \eta_a D_a)^{-1} J_a^\top [y - f(\vartheta_a)].$$

In the simplest case, $D_a = I_p$, i.e. blue, the identity matrix. It should be noted that in any case, in correspondence with a solution, condition $J^\top [y - f(\vartheta)] = 0$ is satisfied.

In general, all iterative methods control the stopping phase of the algorithm in many ways. These are applied simultaneously involving the last two points in the sequence, ϑ_a and ϑ_{a+1}. The convergence of the procedure is based on stopping criteria that control relative changes of the objective function, $S(\vartheta)$, or the relative change of the norm of the parameter vector ϑ.

7.3 Confidence Intervals and Hypothesis Testing

7.3.1 Exact Inference

Exact inference in a nonlinear regression model $y = f(\vartheta) + \varepsilon$, where $\varepsilon \sim \mathcal{N}_n$ $(0, \sigma^2 I_n)$ and $\vartheta \in \mathbb{R}^p$ has been treated in Hartley (1964). If P, an $n \times n$ symmetric and idempotent matrix of rank p, is a projection matrix, i.e. $P(I - P) = 0$:

$$F_\vartheta = \frac{\varepsilon^\top P \varepsilon / p}{\varepsilon^\top (I - P)\varepsilon / (n - p)} \sim F_{p,n-p},$$

where $\varepsilon = y - f(\vartheta)$, i.e. the ratio F_ϑ is distributed exactly as a Snedecor's F with p and $n - p$ degrees of freedom. A $100(1 - \alpha)\%$ confidence region A for the vector ϑ may be defined as follows:

$$A = \{\vartheta : F_\vartheta \leq F_{\alpha;p,n-p}\},$$

where $F_{\alpha;p,n-p}$ denotes the α quantile of the Snedecor's F distribution with p and $n - p$ degrees of freedom. Hartley (1964) proposed to choose a matrix P, which does not depend on ϑ, such that $(I_n - P)f(\vartheta) \approx 0$, and obtained the approximation

$$F_\vartheta \approx \frac{n - p}{p} \frac{[y - f(\vartheta)]^\top [y - f(\vartheta)] - y^\top (I_n - P)y}{y^\top (I_n - P)y},$$

whose contours in ϑ are equivalent to that of constant likelihood, $\varepsilon(\vartheta)^\top \varepsilon(\vartheta) = c$.

7.3.2 Asymptotic Inference and Linear Approximations

Exact inference for confidence regions relies on strong assumptions about the error distribution. However, some relevant results may be obtained asymptotically under weaker conditions.

In a small neighborhood of ϑ^*, the true unknown value of the parameter, a first-order Taylor approximation of the model $f(\vartheta)$ around ϑ^* holds:

$$f(\vartheta) \approx f(\vartheta^*) + J(\vartheta - \vartheta^*), \tag{7.5}$$

where, now, J is the Jacobian in ϑ^*. The deviance $S(\vartheta)$ may be approximated coherently

$$S(\vartheta) = [y - f(\vartheta)]^T [y - f(\vartheta)] \approx [y - f(\vartheta^*) - J(\vartheta - \vartheta^*)]^T [y - f(\vartheta^*) - J(\vartheta - \vartheta^*)]$$
$$= [z - J(\vartheta - \vartheta^*)]^T [z - J(\vartheta - \vartheta^*)] = (z - J\beta)^T (z - J\beta),$$

where $\varepsilon = y - f(\vartheta^*)$, and $\beta = \vartheta - \vartheta^*$. The deviance $S(\vartheta)$ is minimal if β is:

$$\hat{\beta} = (J^T J)^{-1} J^T \varepsilon.$$

Under regularity conditions, for $n \to +\infty$, $\hat{\vartheta}$ belongs, almost certainly, to any small neighborhood of ϑ^*. It is possible, therefore, to use the following approximation

$$\hat{\vartheta} - \vartheta^* = \beta \approx \hat{\beta} = (J^T J)^{-1} J^T \varepsilon,$$

so, writing Equation (7.5) at $\vartheta = \hat{\vartheta}$ we obtain:

$$f(\hat{\vartheta}) - f(\vartheta^*) \approx J(\hat{\vartheta} - \vartheta^*)$$
$$\approx J(J^T J)^{-1} J^T \varepsilon = P_J \varepsilon, \tag{7.6}$$

and the residual vector is

$$y - f(\hat{\vartheta}) \approx y - f(\vartheta^*) - J(\hat{\vartheta} - \vartheta^*) = \varepsilon - P_J \varepsilon$$
$$= (I_n - P_J)\varepsilon, \tag{7.7}$$

where $P_J = J(J^T J)^{-1} J^T$ is a projection matrix. The statistic $(n - p)s^2 = S(\hat{\vartheta})$ may be approximated as

$$S(\hat{\vartheta}) = [y - f(\hat{\vartheta})]^T [y - f(\hat{\vartheta})] \approx \varepsilon'(I_n - P_J)\varepsilon, \tag{7.8}$$

by noting that

$$[f(\hat{\vartheta}) - f(\vartheta^*)]^T [f(\hat{\vartheta}) - f(\vartheta^*)] \approx (\hat{\vartheta} - \vartheta^*)^T J^T J(\hat{\vartheta} - \vartheta^*) \approx \varepsilon^T P_J \varepsilon,$$

then the difference between deviances is

$$S(\vartheta^*) - S(\hat{\vartheta}) \approx \varepsilon^T \varepsilon - \varepsilon^T (I_n - P_J)\varepsilon = \varepsilon' P_J \varepsilon$$
$$\approx (\hat{\vartheta} - \vartheta^*) J^T J(\hat{\vartheta} - \vartheta^*).$$

If a linear approximation in a neighborhood that includes the two points ϑ^* and $\hat{\vartheta}$ is reasonable, then we can replace J with $\hat{J}$, the Jacobian at estimate $\hat{\vartheta}$ in (7.6)–(7.7) and (7.8). Therefore, under regularity conditions, asymptotically, the following results hold

$$\hat{\vartheta} - \vartheta^* \sim \mathcal{N}_p(0, \sigma^2 C^{-1}), \text{ with } C = J^\top J = J^\top(\vartheta^*)J(\vartheta^*);$$

$$(n-p)s^2/\sigma^2 \approx \varepsilon^\top(I_n - P_J)\varepsilon/\sigma^2 \sim \mathcal{X}^2_{n-p};$$

$\hat{\vartheta}$ and s^2 are stochastically independent;

$$\frac{(S(\vartheta^*) - S(\hat{\vartheta}))/p}{S(\hat{\vartheta})/(n-p)} \approx \frac{(\varepsilon^\top P_J \varepsilon)(n-p)}{\varepsilon^\top(I_n - P_J)\varepsilon p} \sim F_{p,n-p}.$$

By using those results we have, approximately,

$$\frac{(\hat{\vartheta} - \vartheta^*)^\top J^\top J(\hat{\vartheta} - \vartheta^*)}{ps^2} \sim F_{p,n-p},$$

and an approximate $100(1-\alpha)\%$ confidence region for ϑ^* is

$$\left\{ \vartheta^* : (\hat{\vartheta} - \vartheta^*)^\top \hat{J}^\top \hat{J}(\hat{\vartheta} - \vartheta^*) \leq ps^2 F_{\alpha,p,n-p} \right\}.$$

Notice that here J (or $\hat{J}$) plays the same role as the model matrix in linear regression.

Similarly, confidence intervals for the individual components of the vector ϑ^* may be obtained by noting that, given a linear combination $a^\top \vartheta$ of the parameter vector, approximately,

$$T = \frac{a^\top \hat{\vartheta} - a^\top \vartheta^*}{s(a^\top C^{-1}a)^{1/2}} \sim t_{n-p},$$

where t_{n-p} is the t-distribution with $n-p$ degrees of freedom; and then an approximate $100(1-\alpha)\%$ confidence interval for $a^\top \vartheta$ is

$$a^\top \hat{\vartheta} \pm t_{\alpha/2,n-p} s(a^\top C^{-1}a)^{1/2},$$

where $t_{\alpha/2,n-p}$ is the $\alpha/2$ percentile of the t distribution with $n-p$ degrees of freedom and C can be estimated by $\hat{C} = \hat{J}^\top \hat{J}$. For $a = (0, \ldots, 0, 1, 0, \ldots, 0)^\top$, with 1 in rth position, we obtain, in particular,

$$\hat{\vartheta}_r \pm t_{\alpha/2,n-p} s(C^{-1}_{rr})^{1/2},$$

where C^{-1}_{rr} denotes the rth diagonal element of the matrix C^{-1}. The simultaneous use of separate intervals for each component involves undersizing of the overall coefficient $(1-\alpha)$ actually associated with each interval. Although a number of more precise proposals have been developed (e.g. see Goeman and Solari, 2011), the most used correction of the intervals is the one proposed by Bonferroni, which modifies the confidence interval as

$$\hat{\vartheta}_r \pm t^{\alpha/2p}_{n-p} s(C^{-1}_{rr})^{1/2}.$$

7.3.2.1 Prediction Intervals

The asymptotic results obtained by linearization of a nonlinear model, can be used to predict the y at some unobserved value $x = x_0$,

$$y_0 = f(x_0, \vartheta) + \varepsilon_0,$$

where $\varepsilon_0 \sim \mathcal{N}(0, \sigma^2)$ are random errors independent of ε. A reasonable estimate of y_0 can be obtained using the fitted model in correspondence with x_0, $\hat{y}_0 = f(x_0, \hat{\vartheta})$. If we consider large n, $\hat{\vartheta}$ will be sufficiently close to ϑ^* and we may linearize the function by applying a first-order Taylor approximation of $f(x_0, \hat{\vartheta})$:

$$f(x_0, \hat{\vartheta}) \approx f(x_0, \vartheta) + f_0'(\hat{\vartheta} - \vartheta),$$

where

$$f_0' = \left(\frac{\partial f(x_0, \vartheta)}{\partial \vartheta_1}, \frac{\partial f(x_0, \vartheta)}{\partial \vartheta_2}, \dots, \frac{\partial f(x_0, \vartheta)}{\partial \vartheta_p} \right),$$

so that

$$y_0 - \hat{y}_0 \approx y_0 - f(x_0, \vartheta) - f_0'(\hat{\vartheta} - \vartheta) = \varepsilon_0 - f_0'(\hat{\vartheta} - \vartheta).$$

For the independence between ε_0 and $\hat{\vartheta}$, we obtain

$$E(y_0 - \hat{y}_0) \approx E(\varepsilon_0) - f_0'E(\hat{\vartheta} - \vartheta) \approx 0$$

$$V(y_0 - \hat{y}_0) \approx V(\varepsilon_0) + f_0'V(\hat{\vartheta} - \vartheta^*)f_0$$

$$\approx \sigma^2(1 + f_0'(J^\mathsf{T}J)^{-1}f_0) = \sigma^2(1 + v_0),$$

where we called $v_0 = f_0'(J^\mathsf{T}J)^{-1}f_0$. Thus, asymptotically

$$y_0 - \hat{y}_0 \sim \mathcal{N}(0, \sigma^2(1 + v_0)).$$

Since s^2 is independent of y_0 and asymptotically independent of $\hat{\vartheta}$, we have that, s^2 is asymptotically independent of $y_0 - \hat{y}_0$ and in asymptotic terms,

$$\frac{y_0 - \hat{y}_0}{s\sqrt{1 + v_0}} \sim t_{n-p}.$$

An approximated $100(1 - \alpha)\%$ confidence interval for the prediction is, therefore,

$$\hat{y}_0 \pm t_{\alpha/2, n-p} s \left[\left(1 + \hat{f}_0'(\hat{J}^\mathsf{T}\hat{J})^{-1}\hat{f}_0 \right) \right]^{1/2}.$$

References

Jelle J Goeman and Aldo Solari. Multiple testing for exploratory research. *Statistical Science*, 26(4):584–597, 2011.

Herman O Hartley. Exact confidence regions for the parameters in non-linear regression laws. *Biometrika*, 51(3/4):347–353, 1964.

George AF Seber and Chris J Wild. *Nonlinear Regression*. New York: Wiley, 1989.

8

Case Studies

After reading this chapter, you should be able to

- Perform a complete analysis with innovation diffusion models
- Select among different innovation diffusion models
- Use innovation diffusion models for scenario evaluation.

8.1 Introduction

This chapter is dedicated to some case studies that employ most models presented in Chapters 2–6. Some of these cases are based on the analysis of one product or technology only, while others consider more than one time series and deal with competition or collaboration between products and technologies. The last case study considers a situation involving multiple time series. Each of the examples illustrated has some peculiar traits that deserve a suitable modeling choice. Moreover, for each of the cases, there is a specific focus on one substantial aspect calling for attention, such as seasonality, dynamic market potential, competition or collaboration, and exogenous shocks. All the case studies will show how diffusion models can be used to explain the growth dynamics of innovation. In some cases, different models can produce different scenarios, and this may suggest diversified strategic decisions.

8.2 Sales of Smartphones

The first case study refers to sales of smartphones by Oppo, a Chinese consumer electronics manufacturer based in Dongguan, Guandong, whose major product lines include smartphones, smart devices, audio devices, power banks, and other

Innovation Diffusion Models: Theory and Practice, First Edition. Mariangela Guidolin.
© 2024 John Wiley & Sons Ltd. Published 2024 by John Wiley & Sons Ltd.
Companion Website: www.wiley.com/go/innovationdiffusionmodels/

electronic products. The brand name Oppo was registered in China in 2001 and launched in 2004. Since then, the company has expanded to 50 countries. According to Wikipedia (https://en.wikipedia.org/wiki/Oppo), in mid-2016, Oppo became the largest smartphone manufacturer in China, selling its phones at more than 200,000 retailers. Oppo was the top smartphone brand in China in 2019 and was ranked to be number five worldwide, according to its market share. Oppo has significantly increased the production and sales of its smartphone lineup over the past five years, shipping 24.7 million units in the second quarter of 2022. Despite considerable growth over the past few years, the total smartphone units sold by Oppo in the second quarter of 2022 were over eight million units fewer than those shipped in the same quarter of 2021. As already seen for other brands in Chapters 2–4, we may consider the series of smartphones produced by a specific manufacturer as being characterized by a life cycle, which may be described with the diffusion models illustrated in Chapters 2–4 of this book. The time series here considered refers to quarterly sales, in million units, at the world level from the first quarter of 2015 to the second quarter of 2022 and is available on the website www.idc.com. Although Oppo started its business well before 2015, we may still consider the first quarter of 2015 as a reasonable starting point for the diffusion process at the world level. By inspecting the plot of quarterly sales in Figure 8.1a, it may be observed that after strong growth of sales in years

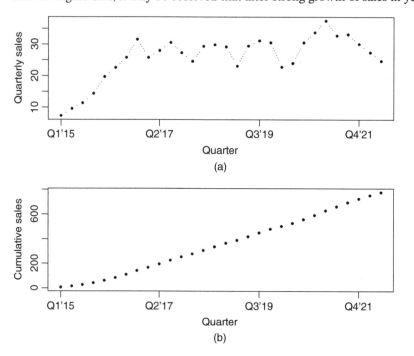

Figure 8.1 Oppo. Quarterly (a) and cumulative (b) sales from 2015 to 2022 (million units).

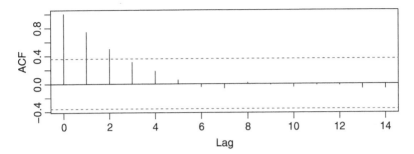

Figure 8.2 ACF for Oppo smartphones, quarterly sales.

2015 and 2016, these have started to show an essentially constant trend, with the possible presence of some seasonal component, although the structure of such seasonality seems to change over time. Specifically, the years 2020 and 2021 exhibit a peculiar behavior, which may be possibly correlated to some effects of the Covid-19 pandemic. Figure 8.1b shows the corresponding cumulative sales. Because quarterly sales show this particular behavior, especially in terms of the presence or not of seasonality, it may be useful to also analyze the autocorrelation function of the data, displayed in Figure 8.2.

The analysis of the autocorrelation function of the quarterly sales, suggests that these are characterized by a trend, although not particularly strong, while the autocorrelation at the seasonal lag 4 is not significant, indicating that there does not appear to be any seasonality. In order to describe the growth in sales of Oppo smartphones, the first model employed, as a baseline, is the Bass Model, even if it may be supposed by the only observation of data plot that this model will not be the most suitable solution for this diffusion process. However, the estimation of the BM shows good results in terms of goodness of fit and parameter estimates: the R^2 is equal to 0.9994, revealing a good fit of the model to the data, but with some possibility of improvement, particularly in the early phase. The results are reported in Table 8.1 and Figure 8.3.

Parameters p and q show a quite peculiar situation since they have essentially the same magnitude, $p = 0.012$ and $q = 0.083$: this is a rare case, because normally parameter p is much smaller than q, suggesting a dominance of the imitation effect

Table 8.1 Parameter estimates of BM for Oppo smartphones.

Parameter	Estimate	s.e.	Lower c.i.	Upper c.i.
m	1153.2	79.8	996.8	1309.5
p	0.012	0.0005	0.011	0.013
q	0.08	0.007	0.06	0.09

$R^2 = 0.9994$.

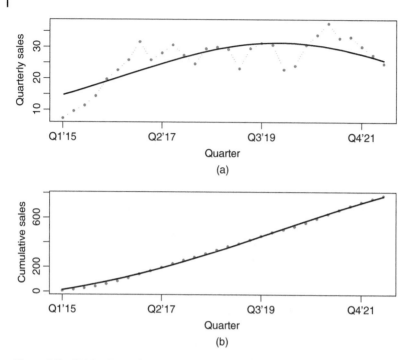

Figure 8.3 BM for Oppo. Quarterly sales (a) and cumulative sales (b).

over the innovative one. However, these values appear to be coherent with the specific structure of the data, with an intense growth at early stages, justifying the high level of p, and a stable trend afterward, which is coherent with the not especially high value of q. Moreover, if we perform an out-of-sample prediction, we may analyze the predicted life cycle for Oppo smartphones according to the standard BM, as shown in Figure 8.4. Following the out-of-sample prediction performed with the BM, the life cycle has just overtaken the maximum peak and is in its declining phase. Overall, the BM seems to provide an acceptable representation of the mean trajectory of the diffusion; however, it clearly overestimates the first part of the series, does not capture the slowdown observed in sales between the end of 2016 and the end of 2020 and predicts a decline in sales that may appear unlikely. Because of these observed flaws, we propose more suitable modeling in order to improve the results.

The first possibility we consider is to try to describe the slowdown in sales through a GBM with one rectangular shock, the results of which are reported in Table 8.2 and Figure 8.5. Overall, the GBM with one rectangular shock implies a great improvement in the goodness of fit, with $R^2 = 0.9999$. The comparison with the BM through the $\tilde{R}^2$ confirms the suitability of the GBM since $\tilde{R}^2 = 0.83$.

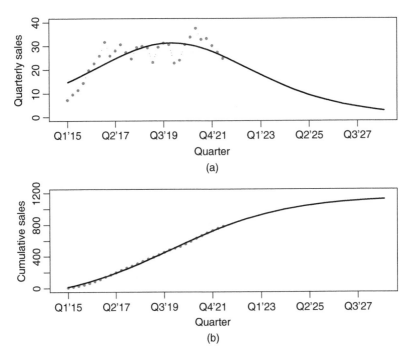

Figure 8.4 Out-of-sample BM for Oppo. Quarterly sales (a) and cumulative sales (b).

Table 8.2 Parameter estimates of GBM with one rectangular shock for Oppo.

Parameter	Estimate	s.e.	Lower c.i.	Upper c.i.
m	910.3	12.7	885.4	935.2
p	0.009	0.0003	0.008	0.010
q	0.18	0.008	0.16	0.19
a_1	11.0	0.25	10.5	11.5
b_1	23.6	0.2335	23.1	24.1
c_1	−0.36	0.02	−0.39	−0.32

$R^2 = 0.9999$.

Moreover, all the estimated parameters are highly significant and suggest interesting insights. First of all, it may be noticed that the market potential is estimated to be smaller with respect to that of the BM, $m = 910.3$, so according to the GBM, the life cycle of Oppo smartphones is shorter than the one predicted with

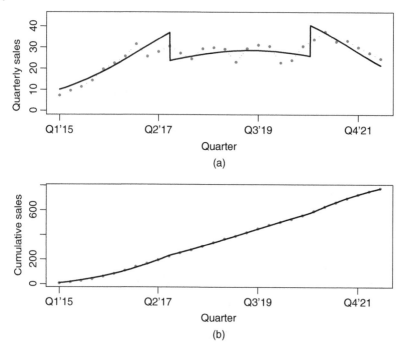

Figure 8.5 GBM with one rectangular shock for Oppo. Quarterly sales (a) and cumulative sales (b).

the BM, where $m = 1153.2$. The parameters of the rectangular shock describe a perturbation with negative intensity $c_1 = -0.36$, starting in the third quarter of 2017, $a_1 = 11$, and ending between the third and the fourth quarters of 2020, $b_1 = 23.6$, with a resumption in sales at the beginning of 2021. Following the description provided by the GBM, the series of OPPO appears to be characterized by three different phases, namely a first growth until the beginning of 2017, a market depression between 2017 and 2020, and a recovery starting in 2021. However, such recovery seems to be very short, because the model soon predicts a decline in sales. The long-term prediction of Oppo sales may be performed also with the GBM and displayed in Figure 8.6, where the presence of the three phases appears even clearer.

A somewhat different and perhaps more reasonable perspective on the Oppo series is obtained if we fit a Guseo–Guidolin model (GGM), which has proved to be able to describe different diffusion patterns in a flexible way. As in the case of the GBM, also, the GGM obtains satisfactory results, shown in Table 8.3 and Figure 8.7. Again, the R^2 has increased with respect to the BM and all the estimated parameters are significant. The comparison with the BM performed through the

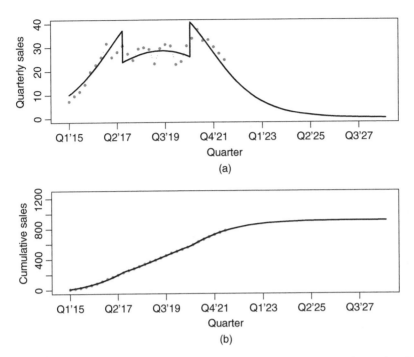

Figure 8.6 Out-of-sample GBM with one rectangular shock for Oppo. Quarterly sales (a) and cumulative sales (b).

Table 8.3 Parameter estimates of GGM for Oppo smartphones.

Parameter	Estimate	s.e.	Lower c.i.	Upper c.i.
K	1269.0	112.4	1048.8	1489.3
p_c	0.0020	0.0002	0.0016	0.0022
q_c	0.12	0.009	0.10	0.14
p_s	0.065	0.0059	0.053	0.077
q_s	0.24	0.037	0.16	0.31

$R^2 = 0.9998$.

$\tilde{R}^2$ confirms that also GGM could be a viable solution for modeling this process, since $\tilde{R}^2 = 0.67$, (It is worth observing that according to the $\tilde{R}^2$, the GBM is a better option: however, goodness of fit indexes are just one mean for model selection and other criteria could be used for the final choice of the model). The estimated parameters are $p_c = 0.002$, $q_c = 0.12$, $p_s = 0.06$, and $q_s = 0.24$, suggesting a clear dominance of imitation in both phases, communication and adoption. The

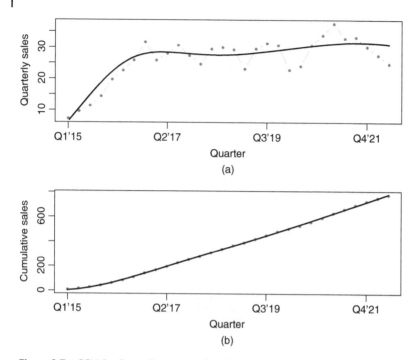

Figure 8.7 GGM for Oppo. Quarterly sales (a) and cumulative sales (b).

graphical display of the predicted values with respect to the observed series well shows the goodness of fit of the GGM: the model is able to describe in a very precise way the early stages of the sales process and captures the slowdown in the period between the end of 2016 and the end of 2020, with a resumption of sales in the last two years, 2020–21. If we make the out-of-sample prediction also for the GGM, we obtain the trajectory represented in Figure 8.8, which makes us appreciate that this model predicts a longer life cycle for Oppo smartphones. This is also testified by the value of the estimated asymptotic market potential $K = 1269$, which is higher than that obtained with the BM and the GBM.

A graphical comparison between the three proposed models, BM, GBM, and GGM is provided in Figure 8.9, clearly showing the difference between them. Figure 8.10 compares the three models in terms of the out-of-sample prediction and allows us to better understand the different perspectives given to market evolution. While it is clear that the BM cannot be considered a good model for this case study, both the GBM and the GGM offer reasonable modeling. However, the GBM predicts a rapid decline in sales and a substantial market saturation in two years, whereas the GGM models the process in a smoother way and implies a more optimistic view of the future sales of Oppo smartphones. In this case,

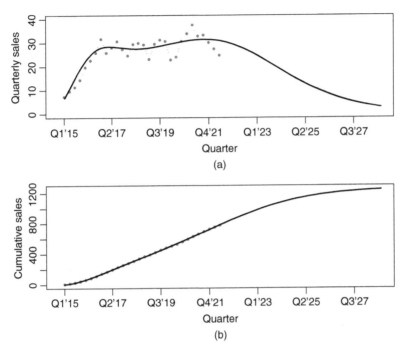

Figure 8.8 Out-of-sample GGM for Oppo. Quarterly sales (a) and cumulative sales (b).

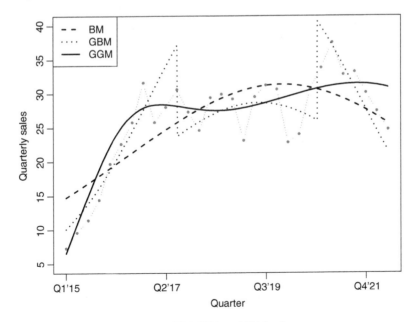

Figure 8.9 Model comparison. GGM, GBM, and BM for Oppo.

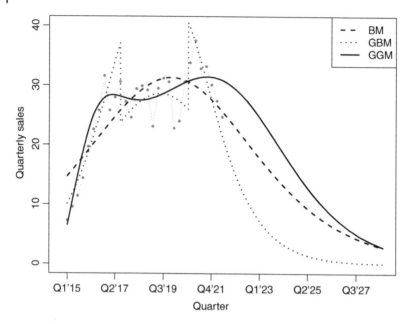

Figure 8.10 Model comparison. Out-of-sample GGM, GBM, and BM for Oppo.

the model forecasts market saturation by the end of 2027. If one had to choose between the GBM and the GGM, some market considerations could orient the selection, depending on whether a pessimistic or an optimistic scenario on the evolution of sales are more advisable.

In any case, the analysis can be completed by studying the autocorrelation function of residuals of the GBM and GGM, which are displayed in Figure 8.11. As is evident, the residual autocorrelation of the GBM has just the first lag significant, while that of the GGM, beside the autocorrelation at lag 1, has two other significant autocorrelations at lags 3 and 4, suggesting the presence of stronger residual structure in the data. To capture this structure, an Autoregressive Moving-Average with external regressor (ARMAX) refinement is applied. The results are reported in Table 8.4 and Figure 8.12, where the improvement obtained can be easily observed. The ARMAX refinement is able to capture the residual variability in the data and, as desired, the parameter λ takes a value very close to 1, $\lambda = 0.9967$, which confirms the GGM as a suitable model for the mean behavior of the data.

8.2.1 Recap

The case study on Oppo sales has been useful to discuss four major points, concerning model selection, out-of-sample prediction, model choice through scenario

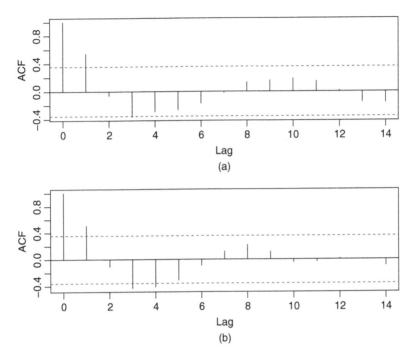

Figure 8.11 Residual ACF after GBM (a) and GGM (b) for Oppo.

Table 8.4 ARMAX refinement for Oppo.

Parameter	ma1	ma2	Intercept	λ
Estimate	1.0524	0.5428	1.2335	0.9967
s.e.	0.1618	0.1698	1.8574	0.0042

$AIC = 146.57.$

considerations, and ARMAX refinement after the model has been selected. This example clearly shows that there may be more than one model proving to be a reasonable choice for describing the data. Model selection can be performed not only through the goodness of fit and other performance measures, but also on the basis of market knowledge and strategic considerations.

8.3 Music Industry in the US

The second case study proposes an analysis of the evolution of the music industry in the United States, by considering the technologies that over time allowed

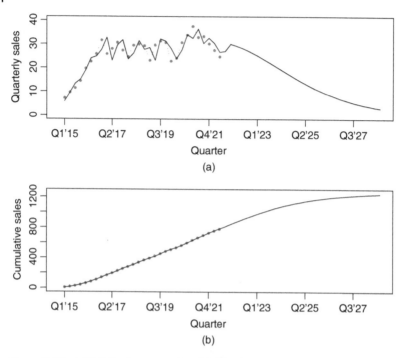

Figure 8.12 SARMAX refinement after GGM for Oppo. already seen for other brands in the previous.

music listening, such as LPs, audiocassettes, and compact discs. The data come from the Recording Industry Association of America (www.riaa.com) and for the purposes of the study, we will consider the yearly shipments of audiocassettes and compact discs (CDs). We do not consider the case of LPs because the data available are incomplete. On the other hand, both audiocassettes and CDs have a complete time series, from the launch in the market until the end of their life cycle. These two technologies had a revolutionary impact on the music industry, modifying the way music was listened to and producing a great increase in the music market. After CDs music consumption radically changed with the diffusion of web-based services: in fact, the rise of the Internet from the end of the 1990s has played a crucial role in the music industry around the world, with streaming and digital services and the rise of mp3 technology. As described, for example, in Guidolin and Guseo (2015), audiocassettes, a technology originally launched in Germany in the mid-60s, were defined as the most successful innovation in the history of audio consumer products because, thanks to its technological characteristics, it was able to make the experience of music consumption mobile and individual. Cassettes realized something that was not possible with LPs, which is the opportunity to enjoy music at almost any time and any place (Guidolin and Guseo, 2015). Indeed, during the 1980s, the great success of cassettes was related to the parallel diffusion of portable pocket recorders and hi-fi players, such as the technology of Walkman, launched in 1979. The Walkman became a standard and dominant technology in the portable music market during the 1980s, like the CD player in the 1990s (Guidolin and Guseo, 2015). Historically, the market peak of prerecorded cassettes was reached in the late 1980s, and sales were overtaken by those of CDs during the early 1990s. Cassettes remained popular for specific applications such as car audio until the early 2000s when the CD player rapidly replaced the cassette player also in new cars. Audio CDs and audio CD players started to be commercially available in Europe and North America in 1983. The growth of CDs gave rise to the digital audio revolution, being first adopted by particular categories of consumers, especially those interested in new music technologies, while the mass market for pop and rock music was reached later when the price of CD players started to decrease. Sales of CDs reached their peak in 2000 and have declined since then, with the parallel advent of music downloads from the Internet. We first start by analyzing the time series of annual sales of audiocassettes, in million units, from 1973 to 2008, which is displayed in Figure 8.13a, while the corresponding cumulative series is in

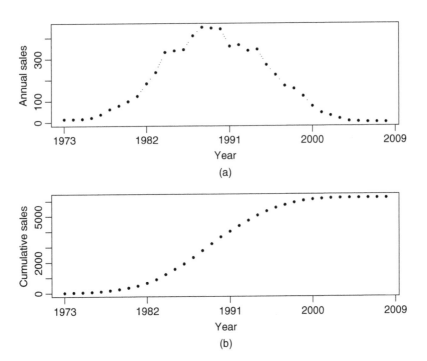

Figure 8.13 Cassettes. Annual sales from 1973 to 2008 (a) and cumulative sales (b) in the US (million units).

Table 8.5 Parameter estimates of BM for cassettes.

Parameter	Estimate	s.e.	Lower c.i.	Upper c.i.
m	6270.3	17.11	6236.7	6303.8
p	0.0021	0.0001	0.0020	0.0023
q	0.29	0.0034	0.28	0.29

$R^2 = 0.9999$.

Figure 8.13b. It is evident that cassettes are a technology with a complete life cycle, showing that its diffusion has essentially followed a Bass-like behavior. Indeed, by inspecting the data, there appears to be no perturbation in this diffusion process, except for some ups and downs around the maximum peak. However, this example seems an excellent application for the simple Bass model.

The results of the application of the BM are presented in Table 8.5 and Figure 8.14. The BM obtains an $R^2 = 0.9999$ and all parameter estimates are significant. Specifically, the values of parameters, $p = 0.0021$ and $q = 0.29$,

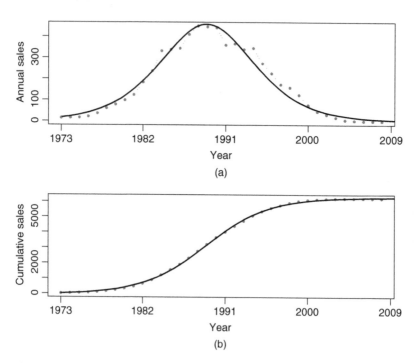

Figure 8.14 BM for cassettes in the US. Annual sales (a) and cumulative sales (b).

suggest a typical case for a BM, where the imitation parameter is a hundred times larger than the corresponding innovation parameter. The market potential m is estimated equal to $m = 6270.3$ (million units). The predictions performed according to the BM appear adequate in describing the evolution of this technology. Given that the complete series is available for this life cycle, we may usefully employ it to perform a prediction exercise, by showing how the trajectory predicted from the model changes depending on the amount of data used. If only 25% of the data is used—in this case nine time points—there is a clear underestimation of the life cycle, if 50% is used the predicted trajectory is close to the one with the entire dataset but some underestimation is still present, while the model fitted with 75% of data is essentially the same of the complete one with the entire time series. This well-shows that the amount of data available has a clear impact on the performance of the model and its predictive ability and reliability. Table 8.6 summarizes the parameter estimates obtained with the four BMs: what may be noticed is the change in the size of the market potential, which is much smaller if just the first nine observations of the series are employed. Instead, as the process reaches its maximum peak, the estimate of the market potential becomes much closer to the one obtained with the entire dataset. Concerning parameters p and q, they are estimated to be much larger if just 25% of the data is used, $p = 0.0042$ and $q = 0.38$, coherently with the fact that in the very early stages of the diffusion process, the growth has an explosive behavior. This exercise is displayed in Figures 8.15 and 8.16, where it is possible to appreciate the difference in the predicted trajectory when different portions of observations are employed. This example also shows that the possibility to well estimate the model not only depends on the number of data employed, but also on the phase of the diffusion process: no matter how many observations one has, if the maximum peak is close or has been already overtaken, this will imply good estimation results. On the other hand, being too far from the peak results in great uncertainties and likely underestimation of the process.

Table 8.6 Parameter estimates of BM for cassettes with 100%, 75%, 50%, and 25% of data.

	100%		75%		50%		25%	
Parameter	Estimate	(s.e.)	Estimate	(s.e.)	Estimate	(s.e.)	Estimate	(s.e.)
m	6270.3	(17.1)	6289.0	(45.6)	5339.5	(132.5)	1831.5	(719.2)
p	0.0021	(0.0001)	0.0022	(0.0001)	0.0018	(0.0001)	0.0042	(0.0013)
q	0.29	(0.0034)	0.28	(0.0048)	0.33	(0.0064)	0.38	(0.0329)
R^2	0.9999		0.9998		0.9998		0.9995	

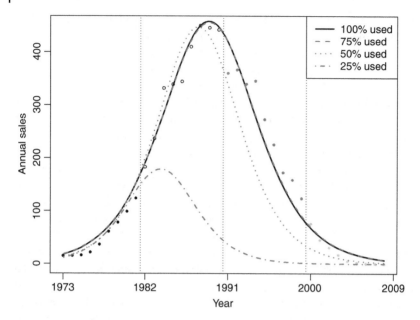

Figure 8.15 BM for cassettes in the US with 100%, 75%, 50%, and 25% of data. Annual sales.

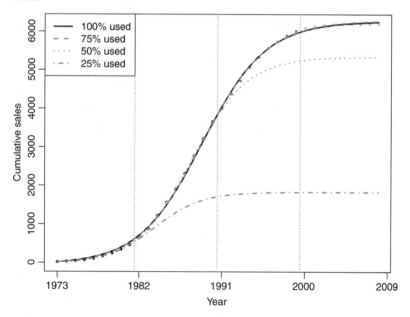

Figure 8.16 BM for cassettes in the US with 100%, 75%, 50%, and 25% of data. Cumulative sales.

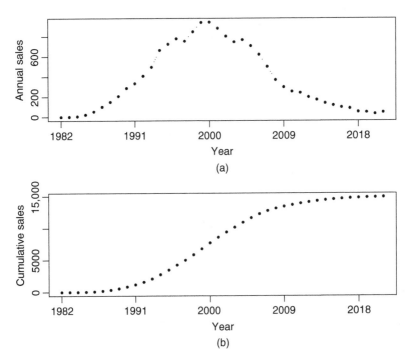

Figure 8.17 Compact discs. Annual sales from 1982 to 2021 (a) and cumulative sales (b) in the US (million units).

Sales of CDs, in million units, from 1982 to 2021, are displayed in Figure 8.17 and show very similar behavior with respect to that of cassettes. The life cycle is essentially complete so that the application of the BM is also proposed for the CDs time series, and reported in Table 8.7 and Figure 8.18. In this case, $R^2 = 0.9998$, and all parameter estimates are significant. The innovation and imitation coefficients are, respectively, $p = 0.0022$ and $q = 0.25$: it may be noticed that with respect to the case of cassettes, the innovative component of the process has remained essentially stable while the imitative effect has slightly decreased. However, what has

Table 8.7 Parameter estimates of BM for compact discs.

Parameter	Estimate	s.e.	Lower c.i.	Upper c.i.
m	14,814.0	49.6	14,716.7	14,911.2
p	0.0022	0.0001	0.0020	0.0024
q	0.25	0.003	0.24	0.26

$R^2 = 0.9998$.

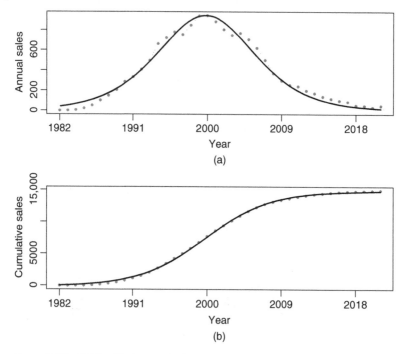

Figure 8.18 BM for CDs in the US. Annual sales (a) and cumulative sales (b).

changed in a dramatic way is the estimated market potential, which has grown to $m = 14,814$ (million units), suggesting that the technological progress from cassettes to CDs has implied an evident growth of the market for music consumption. Obviously, such growth in the market is evident by just observing the data, however, it may be interesting to understand whether the growth of CDs has occurred at the expense of cassettes. Figure 8.19 shows the two series starting from 1982 when CDs entered the market. The data seem to suggest that there has been a technological substitution and to better capture this aspect, we may study the interaction between the two life cycles with a competition model, namely the UCRCD. The results of this application confirm and update some of the findings reported in Guidolin and Guseo (2015).

After performing some comparison, the model selected is a UCRCD with $\delta \neq \gamma$, whose estimation produces very satisfactory results in terms of goodness of fit and parameter estimates, as evidenced in Table 8.8 and Figure 8.20, which provides a focus on the competition phase, after 1982.

A careful inspection of the results suggests some interesting comments. Parameter estimates of the stand-alone phase of cassettes are based on the first nine observations and are clearly the same as those reported in Table 8.6 when 25%

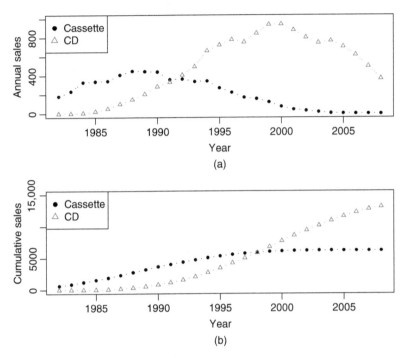

Figure 8.19 Annual sales of cassettes and compact discs from 1982 to 2021 (a) and cumulative sales (b) in the US.

Table 8.8 Parameter estimates of UCRCD for cassettes and compact discs in the US.

Parameter	Estimate	s.e.	Lower c.i.	Upper c.i.
m_a	1831.5	719.2	421.9	3241.1
p_{1a}	0.004	0.0013	0.001	0.006
q_{1a}	0.38	0.032	0.31	0.44
m_c	21,607.5	170.3	21,273.6	21,941.3
p_{1c}	0.008	0.0011	0.006	0.010
q_{1c}	−0.08	0.007	−0.09	−0.06
q_2	0.30	0.012	0.27	0.32
δ	0.19	0.016	0.16	0.22
γ	0.27	0.023	0.22	0.31

$R^2 = 0.9898$.

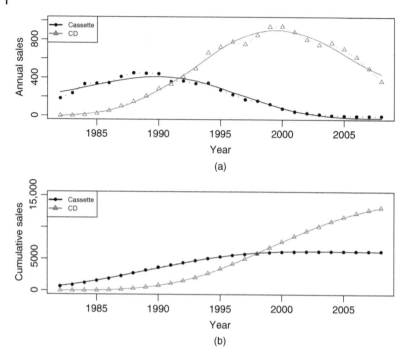

Figure 8.20 UCRCD for cassettes and compact discs. Annual sales (a) and cumulative sales (b).

of data is used. Despite the already mentioned underestimation problems, we may say that they offer a realistic description of the ongoing diffusion process. In the competition phase, i.e. after 1982, the estimated market potential appears to be in perfect agreement with the estimated market potentials of the two univariate processes because $m_c = 21,607$, i.e. close to the sum of the two market potentials estimated with the simple BMs. Referring to the estimated parameters of the competition phase, we may first observe the value of the innovation parameter, $p_{1c} = 0.008$, for cassettes while parameter p_2 has not been reported because non-significant. Therefore while the life cycle of cassettes has been characterized by a positive innovative component, for CDs this has been negligible. This aspect may be probably better understood by also commenting on the values of the imitative components, described in Tables 8.9 and 8.10: the within imitation coefficients,

Table 8.9 UCRCD: Within imitation effects.

Cassettes	$q_{1c} + \delta$	0.11
Compact discs	q_2	0.29

Table 8.10   UCRCD: Cross imitation effects.

Cassettes	q_{1c}	−0.08
Compact discs	$q_2 - \gamma$	0.03

$q_{1c} + \delta = 0.11$, and $q_2 = 0.29$ confirm that the growth of both technologies has been driven by internal forces, while the cross imitation coefficients, $q_{1c} = -0.08$, and $q_2 - \gamma = 0.03$, highlight the presence of a competition-collaboration situation, with CDs competing with cassettes and cassettes reinforcing the growth of CDs. Such an asymmetric relationship, where cassettes substantially aided the rise of CDs, could explain why CDs could enjoy a fast diffusion process in competition, despite the absence of a significant innovation component.

8.3.1   Recap

The case study on the music industry in the US, focusing on historical sales of cassettes and compact discs, has been useful to show the application of the simple BM, for the description of a complete life cycle of a technology. The availability of the complete time series for both technologies has made it possible to perform a worthwhile exercise by estimating the model with truncated data, to show how parameter estimates and predictions can change according to the number of data points employed. In particular, if less than 25% of data is used, the model suffers from an evident underestimation. Besides univariate analyses, the case study has been also employed to test the substitution dynamics between cassettes and compact discs through the UCRCD model, confirming an interesting relationship of competition-collaboration, where the new technology exerts the competitive effect toward the older one while enjoying a sort of collaboration from it.

8.4   Revenues of a Company

In this case study, we analyze the revenues of Twitter, the social network that has been purchased by the entrepreneur Elon Musk in October 2022. Twitter is a real-time microblogging platform, publicly launched in July 2006. When launched, its peculiar characteristic was the limit placed on each post, known as a tweet. Originally, users could only use 140 characters, although that was doubled to 280 in 2017. Developed by former Odeo employees, Twitter realized its first growth of users at the conference SXSW 2007 when the founders showed all the tweets hitting the network in real time. From that moment, Twitter increased its market and reached over 300 million monthly active users. In 2015, it became apparent that growth had slowed, according to Fortune, Business Insider, Marketing Land, and other news websites. In this case study, we focus on the series of quarterly revenues of Twitter, directly available from Twitter (at the website www.statista.com). Although one may question whether the data on revenues may be treated with diffusion models, since revenues are not a product to adopt, we select this example specifically to show that diffusion models may be employed with a larger perspective, in order to describe general growth processes. After all, we can make the reasonable hypothesis that revenues are strictly correlated to the success of the social network, in terms of the increase in users and use, i.e. the diffusion of the social network through the growth of tweets directly affects the parallel increase of revenues. The series of Twitter revenues, displayed in Figure 8.21, goes from the first quarter of 2011 to the second quarter of 2022 (in million US dollars). Although the social network was launched before 2011, we may still consider the available data as a reasonable starting point for the process, because the first observations are quite low with respect to the others. By inspecting the data, both quarterly (a) and cumulative (b), we may notice a clearly growing trend, with a seasonal pattern that starts to be visible at the beginning of 2015. The autocorrelation function of data, in Figure 8.22, confirms

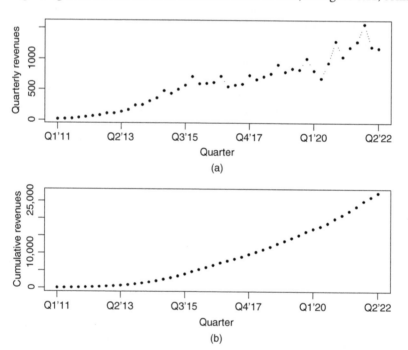

Figure 8.21 Twitter. Quarterly revenues from 2011 to 2022 (a) and cumulative revenues (b).

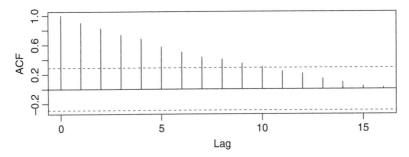

Figure 8.22 ACF for Twitter. Quarterly revenues.

the presence of a growing trend and seasonality since the first 10 autocorrelations are significant.

Also, by observing the data, we may notice that there has been a slowdown in growth between the end of 2015 and the end of 2019, with a resumption of revenues during 2020 and 2021. As already mentioned, such a slowdown in growth had been noticed by various news websites. Given this pattern in data, we may expect that the BM will not be the most suitable to describe the process. Despite this, if we fit a standard BM to the data, we obtain a quite good result, displayed in Table 8.11 and Figure 8.23 with all the parameters being significant. In this case, we may interpret the innovative and imitative components, $p = 0.0019$ and $q = 0.09$, as the direct effect on the revenues of pioneer users of Twitter and of the growth in followers. Although the predicted trajectory according to the BM seems quite adequate, we may notice some typical problems with this model, namely the tendency of overestimating the first part of the trajectory, while underestimating the last portion of data. In fact, according to the BM, revenues have just reached the maximum peak and are starting the decline phase, while the observed time series seems to suggest that the growth is going on.

In order to provide better modeling for this series, we try to fit a GBM, by first considering a rectangular shock, with the purpose of capturing the slowdown in sales that may be observed between the end of 2015 and the first quarters of 2020.

Table 8.11 Parameter estimates of BM for Twitter.

Parameter	Estimate	s.e.	Lower c.i.	Upper c.i.
m	44,633.7	3557.9	37,660.3	51,607.0
p	0.0019	0.0001	0.0018	0.0021
q	0.09	0.004	0.08	0.10

$R^2 = 0.9998$.

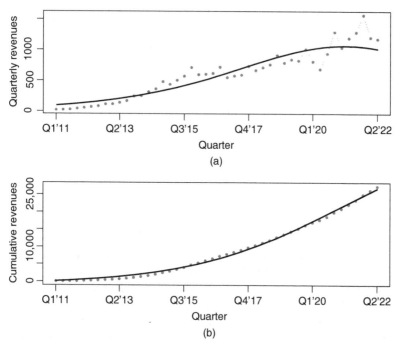

Figure 8.23 BM for Twitter. Quarterly revenues (a) and cumulative revenues (b).

The results of this fitting are summarized in Table 8.12 and Figure 8.24a. Parameter estimates indicate a satisfactory result and the shock appears to be very well captured, between the first quarter of 2016, $a_1 = 21.26$, and the fourth quarter of 2020, $b_1 = 40.6$, with a negative intensity, $c_1 = -0.43$. Despite this good fit, one may notice that after this shock the model predicts a very fast decline in the series,

Table 8.12 Parameter estimates of GBMr1 for Twitter.

Parameter	Estimate	s.e.	Lower c.i.	Upper c.i.
m	35,189.3	492.4	34,224.2	36,154.4
p	0.0008	0.00001	0.0007	0.0009
q	0.17	0.0047	0.16	0.18
a_1	21.26	0.238	20.79	21.73
b_1	40.59	0.150	40.30	40.88
c_1	−0.43	0.011	−0.45	−0.41

$R^2 = 0.9998$.

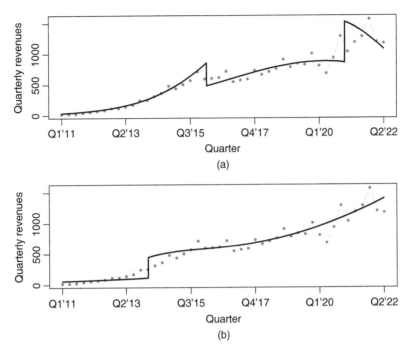

Figure 8.24 GBM with one rectangular shock (a) and one exponential shock (b) for Twitter. Quarterly revenues (million US dollars).

which may appear not realistic in the case of Twitter revenues. This is coherent with the estimated size of the total revenues, $m = 35,189.3$, which is even smaller than the one estimated with the BM, $m = 44,633.7$. In this sense, the GBM with rectangular shock provides a pessimistic scenario for the future growth of Twitter revenues, appearing to be a more suitable solution for making a good description of the data and explaining the slowdown in revenues but not for prediction purposes. To account for this shortcoming, a second option envisages the use of a GBM with exponential shock, giving a different interpretation of the data. Rather than trying to capture the slowdown, the exponential shock has the purpose of describing the acceleration of revenues observed at the very early stages of the series. The estimation of this second model obtains the results presented in Table 8.13 and Figure 8.24b. First of all, the shock is well estimated and suggests the presence of a strong acceleration to growth, $c_1 = 3.01$, started in the first quarter of 2014, $a_1 = 13.05$, but with a negative memory, $b_1 = -0.17$. Unlike in the GBM with rectangular shock, here not all parameters are significant: specifically, the market potential m and the innovation parameter p are nonsignificant, which could suggest that this model is overall less satisfactory than the GBM with rectangular shock, despite having the same goodness of fit, $R^2 = 0.9998$. However, the graphical description of

Table 8.13 Parameter estimates of GBMe1 for Twitter.

Parameter	Estimate	s.e.	Lower c.i.	Upper c.i.
m	320,722	539,910	−737,482	1,378,928
p	0.0002	0.0003	−0.0004	0.0007
q	0.05	0.007	0.03	0.06
a_1	13.05	0.397	12.27	13.83
b_1	−0.17	0.032	−0.23	−0.11
c_1	3.01	0.25	2.52	3.51

$R^2 = 0.9998$.

results is useful to give a motivation for the obtained estimates: unlike the one with rectangular shock, the GBM with exponential shock predicts a strongly increasing trend and this would explain why the estimate of the market potential is unstable and much higher than that of the other two models, $m = 320,722.9$, indicating a strongly increasing process, whose size is difficult to predict. Moreover, the fact of having estimated an exponential shock at the beginning of the series may some-how "hide" the contribution of innovators, i.e. the initializing role of innovators is here in some way substituted by the effect of the exponential shock, which may be interpreted as an external effect, such as an effective advertising campaign, able to stimulate the initial growth of revenues. The fit provided by the GBM with rectangular shock and with exponential shock offers a very different perspective on the growth of Twitter because in the first case, the focus has been on well describing the slowdown, whereas in the second the model has tried to describe the boost in revenues starting in 2014. As already noticed, such a different perspective on the data implies two opposite stories: introducing the rectangular shock causes a rapid decline in revenues, whereas the exponential shock induces continued growth.

Trying to find a sort of compromise between these two views, a third modeling option is offered by the GGM, which obtains a very smooth fit to data as displayed in Table 8.14 and Figure 8.25. Despite some instability in parameter estimates, i.e. K and p_c, the model is able to describe in an almost perfect way the trajectory of observed data, capturing the first growth, the slowdown, and the subsequent resumption in revenues. Parameter estimates clearly indicate a dominance of the imitative component in both the communication and adoption phases, $q_c = 0.11$ and $q_s = 0.21$, with clear word-of-mouth effects. The fact that Twitter revenues are characterized by a dynamic market potential can be interpreted as a sign that the success of the company depends on the parallel increase of the user base. A comparison of the fit for the three proposed models is illustrated in Figure 8.26, from

Table 8.14 Parameter estimates of GGM for Twitter.

Parameter	Estimate	s.e.	Lower c.i.	Upper c.i.
K	188,572.1	263,696.8	−328,264.2	705,408.4
p_c	0.00001	0.00001	−0.00001	0.0001
q_c	0.11	0.004	0.09	0.11
p_s	0.0062	0.0006	0.0050	0.0074
q_s	0.21	0.011	0.19	0.24

$R^2 = 0.9998$.

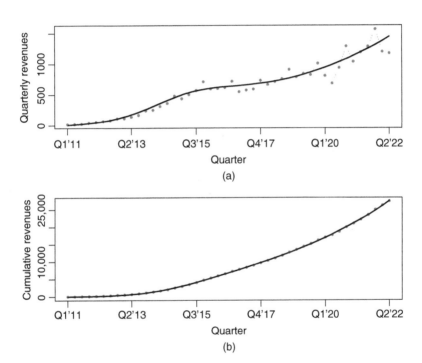

Figure 8.25 GGM for Twitter. Quarterly revenues (a) and cumulative revenues (b).

which the difference between the three models is especially clear. The inspection of residuals of the GGM and their autocorrelation function reported in Figure 8.27 show a significant autocorrelation at lags 5, 6, and 12. In order to account for this residual variability, the analysis is efficiently completed with an ARMAX refinement, ARMA (2,2). The results of the refinement are illustrated in Table 8.15 and Figure 8.28.

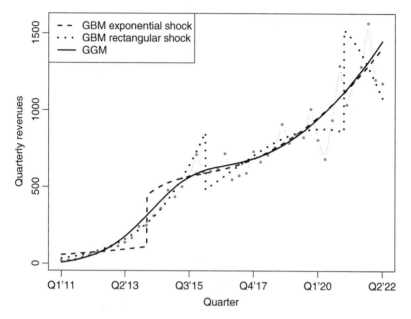

Figure 8.26 Model comparison. GGM, GBM with rectangular shock and with exponential shock for Twitter revenues.

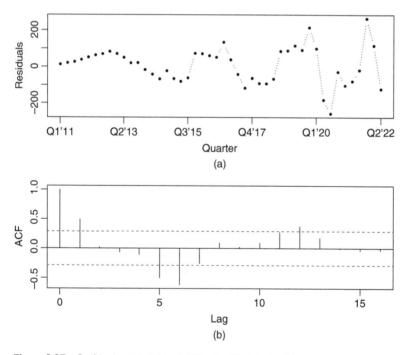

Figure 8.27 Residuals and residual ACF after GGM for Twitter revenues.

Table 8.15 ARMAX refinement for Twitter.

Parameter	ar1	ar2	ma1	ma2	Intercept	λ
Estimate	−0.1144	−0.5205	0.9862	0.9653	19.1259	0.9988
s.e.	0.1597	0.1557	0.1721	0.3128	24.7559	0.0020

$AIC = 531.98$.

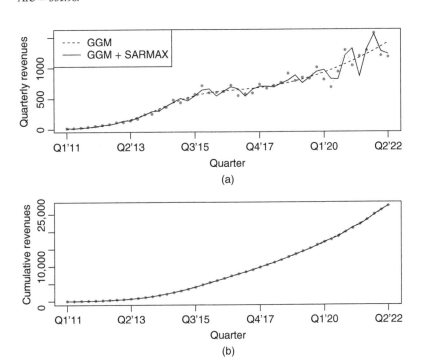

(a)

(b)

Figure 8.28 ARMAX refinement after GGM for Twitter revenues.

8.4.1 Recap

The case study concerning the revenues of a popular social network like Twitter has been proposed to show that diffusion models may be also applied to data that do not concern sales or adoptions of a product. The growth in revenues can be also modeled through diffusion models, as long as there is a clear connection with the growth of the service or product, as in the Twitter case. For the series analyzed, different modeling options have been proposed, showing that the choice of the final model may produce different prediction scenarios, favoring pessimistic or optimistic views on the future evolution of the process. This highlights that the choice of a model is always driven by reasonable assumptions or hypotheses on the evolution of a particular product, technology, or market. A specific understanding of the economic or marketing context should always be combined with statistical modeling.

8.5 The Life Cycle of Tablets

In Chapters 2–4, the innovation diffusion models presented have been employed on the time series of sales of smartphones, computers, and other devices such as Apple iPods. The ICT sector has been particularly innovative in recent years and this explains why it is easy to find examples of innovation life cycles in these markets. Among the various products that have been launched in the last decade, there are tablet computers, a new technology whose purpose has been to combine the features and advantages of laptop computers and telephones. The first company to introduce tablets has been Apple, launching the iPad in 2010, a pioneering technology then followed by several other companies, such as Samsung, Microsoft, and Amazon. This case study considers the time series of sales of the Apple iPad, still the leader in the tablet market. Quarterly data at the world level are displayed in Figure 8.29, along with the corresponding cumulative series, starting from the

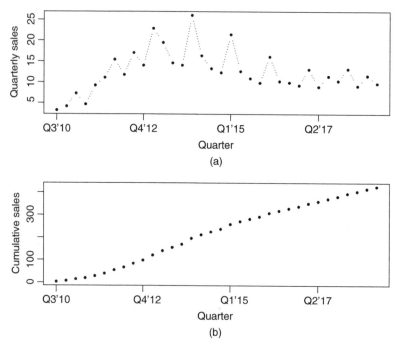

Figure 8.29 iPad. Quarterly sales from 2010 to 2018 (a) and cumulative sales (b).

third quarter of 2010 to the fourth quarter of 2018 (available at www.apple.com). As for other ICT technologies analyzed in Chapters 2–4, we can observe a clear nonlinear trend and a strong seasonality, which confirms that the application of diffusion models should be suitable also in this example. On the other hand, a simple inspection of the behavior of data suggests that the life cycle of this technology seems to be characterized by an asymmetric structure, with a right tail of slowly declining sales starting from the end of 2016. If a BM is employed to describe these data, the overall results appear quite good, as shown in Table 8.16 and Figure 8.30. All the parameters are highly significant and the model can capture in a reasonable way the mean behavior of the series. On the other hand, it can be easily noticed a lack of fit in the first and last portion of data points, where the

Table 8.16 Parameter estimates of BM for iPad sales.

Parameter	Estimate	s.e.	Lower c.i.	Upper c.i.
m	449.3	10.1	429.5	469.1
p	0.0139	0.0007	0.0125	0.0154
q	0.12	0.008	0.10	0.13

$R^2 = 0.9990$.

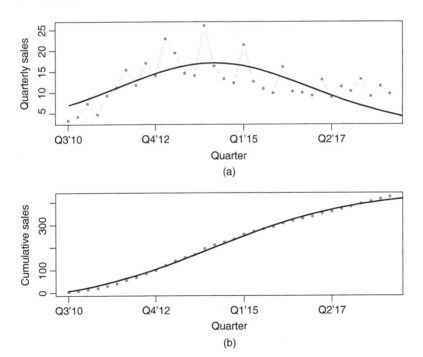

Figure 8.30 BM for iPad. Quarterly sales (a) and cumulative sales (b).

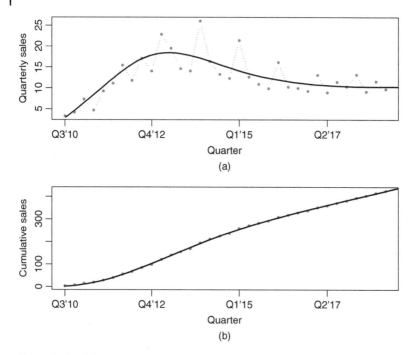

Figure 8.31 GGM for iPad. Quarterly sales (a) and cumulative sales (b).

model, respectively, overestimates and underestimates the observed series, predicting a larger decline in sales than observed in reality. As already seen in other examples in this chapter, this is a typical flaw of the BM, and the GGM may represent a viable solution to account for the nature of the data. The application of the GGM produces an extremely good fit, as can be observed in Figure 8.31, thanks to its ability to describe the first part of the life cycle, characterized by a slow start, the attainment of a maximum peak at the beginning of the year 2013, and the slow decline since then followed by an almost constant level of sales from 2016 on. The fit obtained with the GGM may be usefully compared with that of the BM in Figure 8.32, where the differences between the two model fits can be better appreciated. One may notice that the BM could be a viable model if the data points of years 2017 and 2018 are not considered, whereas the resumption in sales observed from the first quarter of 2017 has changed the behavior of the diffusion demanding different modeling. Table 8.17 shows parameter estimates of the GGM, and it may be seen that not all the parameters are significant, which may suggest some uncertainty regarding the evolution of this life cycle. Despite this potential problem, the GGM may be considered a satisfactory choice for this case because, thanks to its flexibility, it allows fitting data patterns that depart from the classical

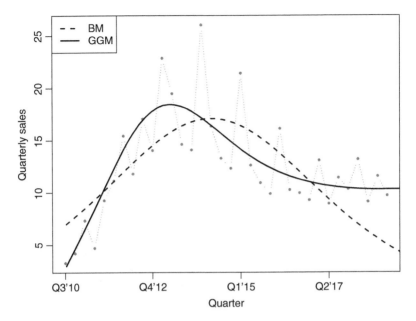

Figure 8.32 Model comparison. GGM and BM for iPad sales.

Table 8.17 Parameter estimates of GGM for iPad sales.

Parameter	Estimate	s.e.	Lower c.i.	Upper c.i.
K	2580.6	34,382.7	$-64,808.4$	69,969.6
p_c	0.0005	0.0122	-0.0234	0.0243
q_c	0.03	0.038	-0.04	0.10
p_s	0.030	0.0023	0.025	0.034
q_s	0.20	0.016	0.16	0.23

$R^2 = 0.9997$.

bell shape of the BM. The life cycle analysis is usually completed with SARMAX modeling, allowing the capture of the seasonality in the data in a quite good way, as shown in Figure 8.33 and Table 8.18.

The Apple iPad example suggests that a different, perhaps new, type of life cycle may be existing, characterized by an asymmetric shape, with fast initial growth and a slow decline, eventually reaching a constant level of sales. In a completely different context, concerning the spread of the Covid-19 pandemic, such an asymmetric diffusion process has been commented on by Guidolin and Manfredi (2023), noting that in most countries the spread of the virus has been

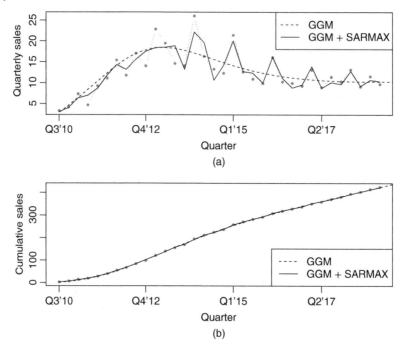

Figure 8.33 SARMAX refinement after GGM for iPad. (a) Quarterly sales, (b) cumulative sales.

Table 8.18 SARMAX refinement for iPad.

Parameter	ar1	ma1	sar1	sar2	sma1	sma2	intercept	λ
Estimate	−0.1200	0.5510	1.1839	−0.1850	−0.3585	−0.5916	0.9088	0.9986
s.e.	0.3631	0.2989	0.3209	0.3199	0.3696	0.3091	1.4257	0.0036

AIC = 139.72.

characterized by rapid growth and a much slower recession, due to some inertial dynamics. This may be the sign that some diffusion processes are modifying their nature and those products and technologies, rather than disappear from the market, could have longer life cycles and exhibit an "endemic" behavior through time, resembling the pattern of the spread of a pathogen.

8.5.1 Recap

The life cycle of the Apple tablet, iPad, shows a peculiar behavior with a visibly asymmetric shape. After the attainment of the maximum, peak sales slowly declined and reached a quite stable trend. To model such behavior in the data, a GGM is found to be the most suitable option because it is able to account for this special behavior of the data, which may suggest the existence of a new type of life cycle, where sales after reaching the maximum peak start to decline slowly and eventually stabilize with a constant trend.

8.6 Energy Transition in Germany

This case study analyzes the energy transition in Germany, by employing diffusion models to predict the evolution of consumption of different energy sources, namely coal, gas, nuclear, and renewables. The case of Germany is especially interesting because, in late 2010, this country started the *Energiewende*, a major plan for making its energy system more efficient, supplied mainly by renewable energy sources. The country has adopted a strategy for an energy pathway to 2050, foreseeing an accelerated phaseout of nuclear power by 2022. In order to achieve the ambitious Energiewende by 2030, there are some clear objectives that need to be realized: half of all electricity supply will come from renewable energy sources and coal use will be phased out by 2038. Germany has been an early leader in wind and solar photovoltaic energy and increased its targets with 20 GW of wind by 2030 and 40 GW by 2040. At this stage, the Energiewende has been mainly developed in electricity generation, where it has had the effect of increasing renewable electricity generation. While coal still represents the largest source of electricity, renewables have almost replaced a large portion of nuclear over the last decade.

In this case study, we adopt the same perspective already described in Chapters 2 and 6, by considering energy sources as technologies that are adopted by a certain market or social system, so diffusion models may represent a suitable modeling approach to describe the energy market evolution in this country. The data considered come from the BP Statistical Review for World Energy 2021, available at www.bp.com, and refer to the time series of consumption expressed in ExaJoule.

For coal, gas, and nuclear the first available data point is 1965, while for renewables the series is shorter, starting in 1992. All the series have been considered until the year 2019, although the 2020 data point is available. However, because 2020 was a unique year due to the Covid-19 pandemic and the subsequent lockdowns, it has been considered more reasonable to take it out of the analysis. Figure 8.34 shows all the series together and is clearly suggestive of a competitive energy market, where different sources are competing with each other, and the growth of one source may occur at the expense of another: for example, the growth of renewables seems to have caused a parallel decrease in nuclear consumption, while the increase in gas consumption could have determined the observed fall in coal consumption. Obviously, the only inspection of data cannot support clear conclusions

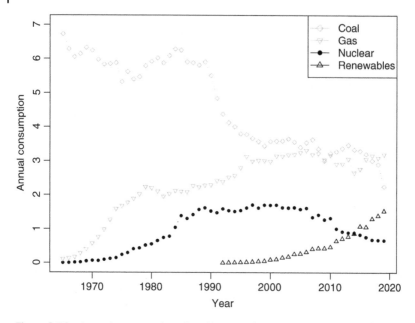

Figure 8.34 Annual consumption of coal, gas, nuclear, and renewables in Germany.

and some more insights can be obtained through models able to test the actual presence of such interactions. We first analyze each energy source individually, by selecting for each time series the most suitable diffusion model—BM, GBM, and GGM—, by also showing some comparisons, whenever reasonable. We then propose a competition modeling between two energy sources, namely nuclear and renewables, by employing the UCRCD model. Finally, the analysis is completed by considering the application of the UCTT model, accounting for the interaction between coal, nuclear, and renewables. The only energy source for which we provide univariate prediction is gas, which is characterized by a peculiar behavior and poses important challenges in modeling, as clarified later in the text.

8.6.1 Coal

The first energy source we consider is coal, whose consumption data, from 1965 to 2019, are shown in Figure 8.35, in both annual and cumulative terms. As is evident by observing the annual series, although being the first energy source for electricity production, the trend is clearly declining and the first available data point in 1965 is evidently not the starting point of coal energy consumption. Despite the particular nature of the data, which may pose some problems in modeling, a careful evaluation suggests performing an analysis with a GBM with one exponential shock, able to capture the mean trajectory in a satisfactory way. Fitting a GBM

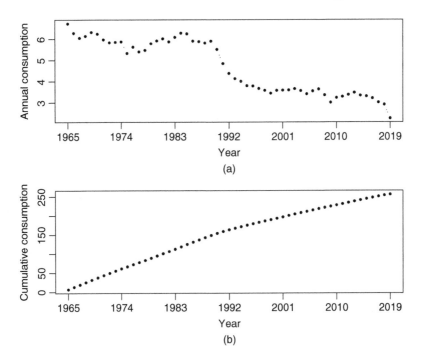

Figure 8.35 Coal energy. Annual consumption in Germany (a) and cumulative consumption (b) (in ExaJoule).

on the coal data produces the results reported in Table 8.19 and Figure 8.36. The graphical description of the model fit shows that the GBM is a very suitable solution for the mean trajectory of coal consumption, even if some residual variability is clearly not described through it. Although some instability in the estimates can

Table 8.19 Parameter estimates of GBM with one exponential shock for coal energy.

Parameter	Estimate	s.e.	Lower c.i.	Upper c.i.
m	973.1	1553.1	−2070.8	4017.2
p	0.006	0.0101	−0.013	0.026
q	0.003	0.013	−0.021	0.029
a_1	27.37	0.3202	26.74	28.00
b_1	0.011	0.0050	0.010	0.020
c_1	−0.31	0.0195	−0.35	−0.27

$R^2 = 0.9999$.

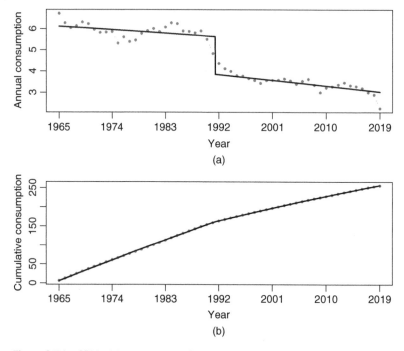

Figure 8.36 GBM with one exponential shock for coal. Annual consumption (a) and cumulative consumption (b).

be noticed, due to the fact that the series is incomplete, the estimated parameters p and q interestingly show that these two components have essentially the same magnitude, $p = 0.006$ and $q = 0.004$ with the imitative effect being especially low and smaller than the innovative one: this may be explained as the sign that the life cycle of coal is in its declining phase, coherently with the country plans of phasingout of this energy source in the next future. Both the graphical inspection of the results and the estimated parameters confirm that the exponential shock is able to describe effectively the fall in consumption observed in 1992, $a_1 = 27.3$. Interestingly, both gas and renewables started to grow significantly in 1992, which may partly explain such a strong decrease in the coal series. Figure 8.37 provides a graphical comparison between the fit obtained with the GBM and the one with a simple BM, showing that also the simple BM can be a fair solution for the series: clearly, such a model cannot capture the strong decrease in consumption observed in 1992, yet it offers a realistic description of the evolution of coal market. By computing a comparison between the BM and the GBM through the $\tilde{R}^2$, we obtain $\tilde{R}^2 = 0.46$, which confirms the relevant role played by the exponential shock in describing the data.

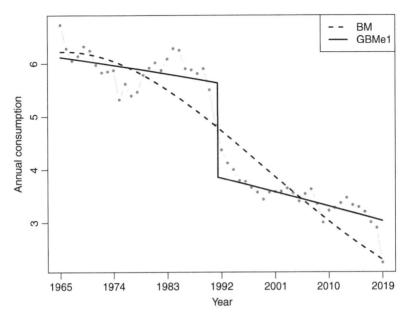

Figure 8.37 Model comparison. GBM with one exponential shock and BM for coal energy in Germany.

8.6.2 Gas

Germany is the largest natural gas market in Europe and gas has been playing an increasingly relevant role in electricity production in Germany, as can be understood by observing the time series of consumption from 1965 to 2019, displayed in Figure 8.38: a rapid growth characterized the 70s with the attainment of a local peak in 1979, while in the 80s, there was a slowdown in consumption until approximately 1992 when consumption restarted. Starting with 2000, the series has shown an essentially stable trend. Considering the specific structure of the data, with a first phase of growth, a slowdown, and a second phase of growth, a GGM has been selected as the most appropriate model. The results are reported in Table 8.20 and Figure 8.39. One could ask what is the rationale for using a model with a dynamic market potential in energy consumption, accounting for the existence of two phases, i.e. communication and adoption. As already observed, energy sources may be considered technologies that need to be accepted by markets and social communities, which implies building a spread consensus around their employment and potential benefits. Especially in the case of energy sources, a market needs to be built over time and be based on the knowledge of potential consumers, so that the GGM appears as a meaningful choice, not only for its flexibility in fitting data but also for its interpretative power. The analysis of parameter estimates

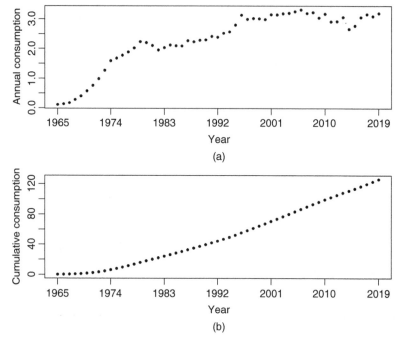

Figure 8.38 Gas energy. Annual consumption in Germany (a) and cumulative consumption (b) (in ExaJoule).

shows that in both the communication and adoption phases, gas consumption has been characterized by a quite strong imitation component, $q_c = 0.11$ and $q_s = 0.23$. In spite of the good results obtained, the fitting illustrated in Figure 8.39 shows an evident model flaw in capturing the last four data points, namely from 2016 to 2019, after the vertical line, corresponding to the year 2016. In fact, the GGM

Table 8.20 Parameter estimates of GGM for gas energy.

Parameter	Estimate	s.e.	Lower c.i.	Upper c.i.
K	152.8	1.25	150.3	155.2
p_c	0.0005	0.00001	0.0005	0.0005
q_c	0.11	0.0012	0.10	0.11
p_s	0.0165	0.0016	0.0133	0.0197
q_s	0.23	0.0133	0.21	0.26

$R^2 = 0.9999$.

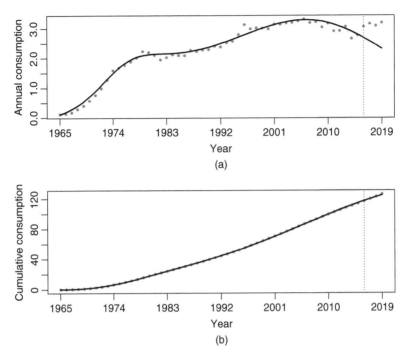

Figure 8.39 GGM for gas. Annual consumption (a) and cumulative consumption (b).

predicts a decline in consumption, whereas the data are clearly growing. The same problem is encountered by fitting a simple BM to the data, as shown in Figure 8.40, where the two models are compared, confirming the better description provided by the GGM, but also the failure of both models in capturing the last portion of data.

The recent growth in gas consumption can be explained by a combination of increased demand in the residential sector, industry, heat, and power generation, as a consequence of the planned phaseout of nuclear and coal power, which has notably strengthened the position of gas in the German electricity system. Because of the observed problems with univariate models in describing the gas series, the story of this energy source should be considered a peculiar and complex case, whose evolution is difficult to predict, especially in light of the perturbations due to the Covid-19 pandemic and the limitations to Russian gas availability started in 2021. Given its complexity, careful management of this case suggests not employing it for multivariate models, in order to capture the possible interaction with other energy sources, although one could expect that such interaction exists, as a simple inspection of the observed data seems to suggest.

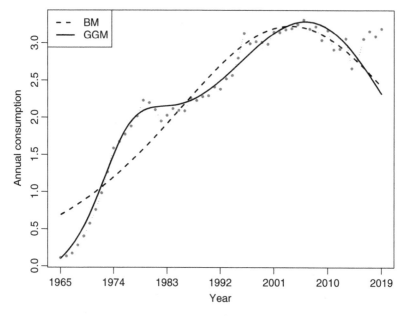

Figure 8.40 Model comparison. GGM and BM for gas energy in Germany.

8.6.3 Nuclear

The decision to phase out all nuclear power by 2022 is one of the central goals of the German energy transition. In 2009, the government decided to support the role of nuclear energy as a "bridging technology" for the transition to renewable energy, and the Fukushima accident in March 2011 reinforced the belief of a necessary elimination of this energy source. This phase-out process is evident by observing the annual consumption data in Figure 8.41. However, it may be also noticed that the decline in consumption started in the early 2000s, well before the Fukushima events. In fact, the consumption of nuclear energy increased until the end of the 80s, and then, starting with the 90s, an almost constant trend has been observed until the first years of the 2000s, when a clear decline began. It may be concluded that more than the Fukushima accident, a general skepticism toward nuclear energy started to arise after the Chornobyl accident, which occurred in 1986, causing a marked slowdown in consumption. The model that most appropriately describes this pattern in the data is the GGM, because of its known capacity to fit effectively different diffusion processes, departing from the bell-shaped form of the BM. The summary of the model fit is reported in Table 8.21 and Figure 8.42, where the good performance of the model can be appreciated both by analyzing parameter estimates and by observing

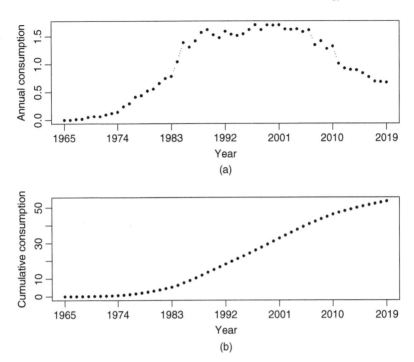

Figure 8.41 Nuclear energy. Annual consumption in Germany (a) and cumulative consumption (b) (in ExaJoule).

Table 8.21 Parameter estimates of GGM for nuclear energy.

Parameter	Estimate	s.e.	Lower c.i.	Upper c.i.
K	57.4	0.205	57.0	57.8
p_c	0.0004	0.00002	0.0004	0.0004
q_c	0.14	0.0019	0.13	0.14
p_s	0.0029	0.0002	0.0025	0.0034
q_s	0.22	0.0063	0.21	0.24

$R^2 = 0.9999$.

the predicted trajectory. As already done for other energy sources, a comparison with a simple BM is displayed in Figure 8.43, where the improvement obtained with the GGM is easily noticed: the BM overestimates the first portion of the data, fails to capture the almost constant behavior of data in the central part of the diffusion process, and slightly underestimates the last part of the observations. The gain implied by the GGM with respect to the BM is considerable, because $\tilde{R}^2 = 0.94$,

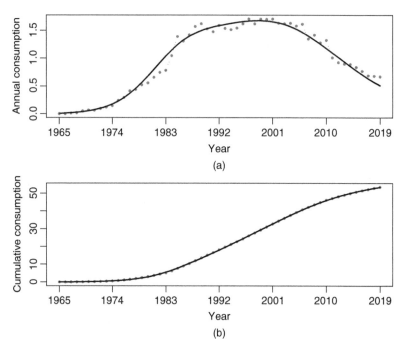

Figure 8.42 GGM for nuclear. Annual consumption (a) and cumulative consumption (b).

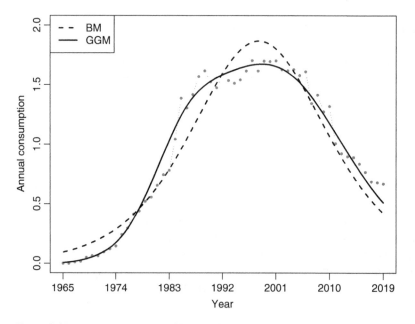

Figure 8.43 Model comparison. GGM and BM for nuclear energy in Germany.

even if it should be recognized that the prediction performed with the BM is not very far from that of the GGM on the last part of the data points. This normally happens when the diffusion process has almost finished, as in this case. As already commented in other case studies, whenever the life cycle is essentially complete the BM can produce a meaningful prediction.

8.6.4 Renewables

Renewable energy is at the center of the German energy transition. In recent years, the share of renewable energy has strongly increased. According to the International Energy Agency (www.iea.org), the most impressive growth has taken place in electricity generation, where renewable energy rose from below 5% in 1998 to 35% in 2018. Although Germany is a leading country in solar photovoltaic power, in recent years, the growth has been dominated by wind power. This country was able to rapidly develop its share of renewable energy in electricity generation thanks to strong policy commitment. In particular, incentive mechanisms based on feed-in tariffs stimulated the adoption of renewable electricity, by supporting the diffusion of wind power since the early 2000s. In the period between 2009 and 2012, Germany registered a rapid diffusion in photovoltaic installations, especially in distributed generation plants in the commercial sector. Such a strong effort has led to 35% of total electricity generation in 2018.

In this section, the time series of consumption of renewables is analyzed by accounting for solar photovoltaic and wind energy together from 1992 to 2019. Figure 8.44 displays the growth dynamics of these energy sources in both annual and cumulative terms, showing a strongly increasing trend, for which diffusion models appear especially suitable. The best model fit for these data is a GBM with one exponential shock, whose estimated parameters and fitting are reported in Table 8.22 and Figure 8.45. The analysis of the parameters suggests some interesting insights. First of all, the estimated market potential $m = 51$ has the same magnitude as that estimated for nuclear energy $K = 57$, indicating that the two sources of energy may be considered direct competitors, at least in terms of market size. However, we may expect that the market for renewables will significantly increase in the future and the estimate for m will become larger with newly available data. It is worth mentioning that, also in this case, the 2020 data point has not been considered, despite being available, because the Covid-19 pandemic and the related economic crisis surely have exerted an effect on these energy sources, at least slowing down their growth. Commenting on the other estimated parameters, a few words could be spent on the very low innovation parameter, $p = 0.0002$, typical of most renewable energy dynamics. This is due to the fragile role of innovators, which well explains the need for incentive measures, able to stimulate an otherwise stagnating market. In fact, the parameters of the estimated exponential

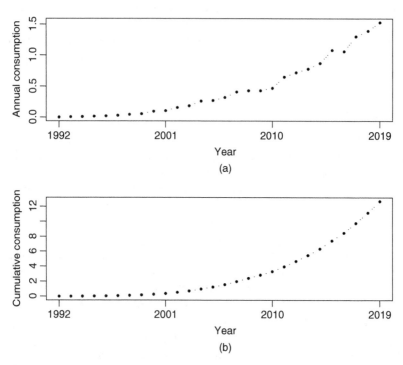

Figure 8.44 Renewable energy. Annual consumption in Germany (a) and cumulative consumption (b) (in ExaJoule).

Table 8.22 Parameter estimates of GBM with one exponential shock for renewable energy.

Parameter	Estimate	s.e.	Lower c.i.	Upper c.i.
m	51.7	8.5108	35.0	68.3
p	0.0002	0.00003	0.0001	0.0003
q	0.16	0.0072	0.15	0.17
a_1	9.14	0.4103	8.34	9.95
b_1	−0.31	0.0625	−0.43	−0.19
c_1	1.77	0.2248	1.33	2.21

$R^2 = 0.9999$.

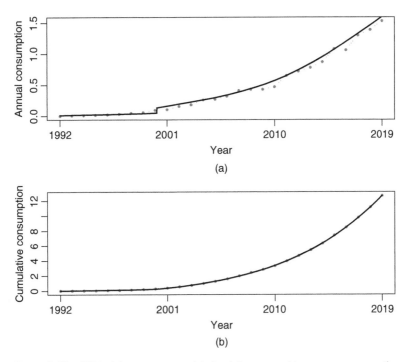

Figure 8.45 GBM with one exponential shock for renewable energy consumption. Annual consumption (a) and cumulative consumption (b).

shock may be interpreted as the positive effect of feed-in tariffs introduced in the early 2000s to foster the adoption of wind and solar power, with $a_1 = 9.14$ saying that the shock started in 2000, with a strong intensity $c_1 = 1.77$, but with a negative memory, $b_1 = -0.31$, which suggests that this boost to growth was deemed to decrease over time, returning to a "normal" situation. The fact that incentives had a limited effect in time is meaningful because they were introduced to stimulate growth but once the process started, renewables needed to stay in the market without constant external support. One could wonder whether introducing the exponential shock was really necessary for modeling the series and whether a simple BM was not enough. As usual, we may perform a comparison by considering the gain in the goodness of fit obtained with the GBM, in this case, $\tilde{R}^2 = 0.86$, which highly confirms the significant improvement, but also the graphical contrast of the two models, displayed in Figure 8.46 provides a clear motivation for preferring the GBM. Even if the BM produces an acceptable overall description of the series, it fails to capture the growing dynamics in the 2019 data point. According to the BM, renewables have just reached the maximum peak of their diffusion process and are going to decline. On the other hand, the GBM predicts a more

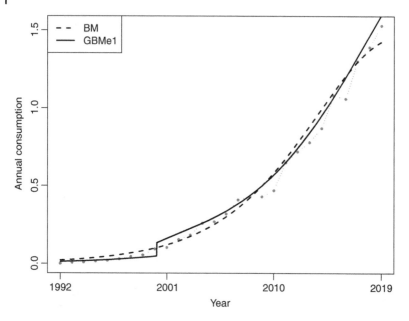

Figure 8.46 Model comparison. GBM with one exponential shock and BM for renewable energy in Germany.

optimistic and reasonable trajectory for the future of these energy sources with continued market expansion. It is evident that such a different future scenario is due to the introduction of the exponential shock, which is able to better describe the first part of the data and allows not to underestimate the last observations as the BM notably tends to do.

8.6.5 Nuclear and Renewables

As already commented, nuclear and renewable energy appear to be direct competitors. Such competition may be inferred by the only observation of data in Figure 8.34, where renewables seem to benefit from the parallel decline of nuclear energy. However, it is interesting to try to verify the concrete existence of this significant interaction and the UCRCD model may be employed to this end. The UCRCD model that best fits the data is the unrestricted model with $\delta \neq \gamma$. The findings of this application are reported in Tables 8.23, 8.24, and 8.25. The graphical display of the model fit is reported in Figure 8.47, making appreciate the extremely satisfactory description of both series. Focusing on parameter estimates, the first thing to observe is the overall stability of the model, with all the parameters being significant, except for parameter p_2, which may be considered almost equal to zero. The estimate of this parameter is nonetheless

Table 8.23 Parameter estimates of UCRCD.

Parameter	Estimate	s.e.	Lower c.i.	Upper c.i.
m_a	26.6	0.73	25.1	28.0
p_{1a}	0.0007	0.00001	0.0006	0.0007
q_{1a}	0.23	0.004	0.22	0.24
m_c	99.9	9.87	80.5	119.2
p_{1c}	0.012	0.0012	0.010	0.014
p_2	0.001	0.0015	−0.002	0.003
q_{1c}	−0.145	0.0150	−0.176	−0.114
q_2	0.342	0.0683	0.208	0.475
δ	0.183	0.0186	0.146	0.219
γ	0.343	0.0730	0.200	0.487

$R^2 = 0.9915$.

Table 8.24 UCRCD. Within imitation effects.

Nuclear	$q_{1c} + \delta$	0.03
Renewables	q_2	0.342

Table 8.25 UCRCD. Cross imitation effects.

Nuclear	q_{1c}	−0.145
Renewables	$q_2 - \gamma$	−0.002

reported because it confirms the already noted weakness of innovators in the diffusion of renewables.

Regarding the other results, it can be noticed that the consumption of nuclear energy is efficiently described with a simple BM, through parameters m_a, p_{1a}, and q_{1a}, until 1992 when renewables entered the market. In this case, the BM for nuclear energy has been estimated by using only the first 27 data points, from 1965 to 1991, producing the good fit observed in Figure 8.47. At this point, one could wonder why here the BM is able to well describe the data, while in 8.6.3 it had a poor performance, and the reason is the different amount of data used. As already seen in Section 8.3, a different number of data points employed for model estimation typically do not produce the same results in terms of parameter estimates and fitting.

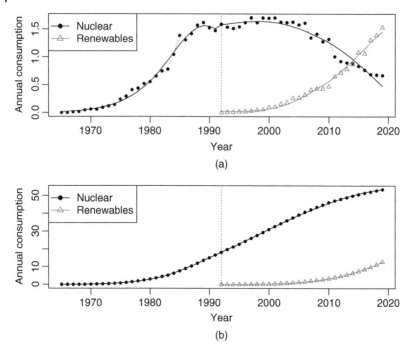

Figure 8.47 UCRCD model for nuclear and renewable energy. Annual consumption (a) and cumulative consumption (b).

Turning the attention to the competition phase, the findings are especially interesting, since they confirm the existence of significant competition between nuclear and renewables. Tables 8.24 and 8.25 help to understand the estimated effects, in terms of within and cross imitation. The within imitation parameters, $q_{1c} + \delta = 0.03$ for nuclear, and $q_2 = 0.342$ for renewables, testify a positive internal dynamics in consumption of both energy sources, although much stronger for renewables. The cross imitation parameters, $q_{1c} = -0.145$ for nuclear and $q_2 - \gamma = -0.002$ for renewables, show that each energy source is exerting a competitive effect on the other, although the effect of renewables on nuclear is much more evident, whereas nuclear energy has a negative but quite limited influence on the growth of renewables. Although one possible problem with the fitted UCRCD is the size of the market potential in competition, m_c, which appears to be underestimated, all these findings appear meaningful and lead to consistent conclusions, suggesting that renewables are playing a relevant role in nuclear phaseout.

8.6.6 Coal, Nuclear, and Renewables

The analysis performed with the UCRCD model may be complemented by the application of the UCTT model, by accounting for the interaction between renewables, nuclear, and coal energy. The use of the UCTT model is typically not straightforward, because of the difficulty of modeling three series of data and managing a quite high number of parameters, so the results proposed in this application are particularly good for the ability of the model to capture simultaneously the trajectories of the three energy sources considered.

As a general note, the model obtains a noteworthy result in terms of goodness of fit $R^2 = 0.9965$ and the significance of almost all the involved parameters, displayed in Table 8.26. In fact, despite some slight instabilities in only two innovation parameters, namely p_2 and p_{3d}, the model is well estimated, and the relationships between the three competitors may receive a clear interpretation. The model fitting is shown in Figures 8.48 and 8.49. Visual inspection of the results shows the effectiveness of the UCTT model to account for these patterns: renewables exhibit an initial exponential consumption trend affecting coal and nuclear trajectories. As already noticed in the univariate analysis, coal consumption is in a clearly declining phase: after an initial period of stability, a sharp decrease starts to be evident around the early 90s, i.e. when renewables entered the market. The model is able to efficiently capture the nuclear series characterized by an increasing trend during

Table 8.26 Parameter estimates of UCTT for Germany energy.

Parameter	Estimate	s.e.	Lower c.i.	Upper c.i.
m_c	222.6	3.6304	215.5	229.7
p_{1c}	0.030	0.0008	0.028	0.031
p_2	0.0004	0.0005	−0.0005	0.0013
q_{1c}	0.96	0.078	0.80	1.11
q_2	0.44	0.037	0.36	0.51
δ	−0.95	0.079	−1.10	−0.79
γ	0.44	0.039	0.36	0.52
m_d	404.0	14.04	376.4	431.5
p_{1d}	0.103	0.0329	0.038	0.167
p_{2d}	0.003	0.0056	0.006	0.008
q_{1d}	0.68	0.2210	0.26	1.09
q_{2d}	0.26	0.051	0.16	0.36
q_3	0.58	0.094	0.39	0.76
ζ	−0.96	0.305	−1.56	−0.36
ρ	0.36	0.070	0.22	0.50
ξ	0.58	0.097	0.39	0.77

$R^2 = 0.9996$.

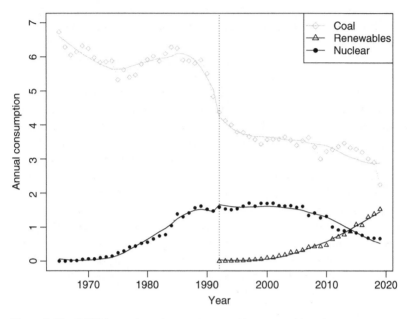

Figure 8.48 UCTT for coal, nuclear, and renewable energy. Annual consumption.

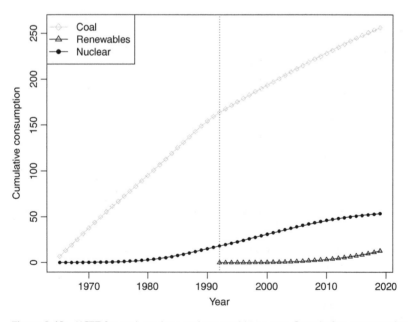

Figure 8.49 UCTT for coal, nuclear, and renewable energy. Cumulative consumption.

the first phase, a more stable behavior in the 80s, and then, similarly to coal, a decline phase corresponding to the growth of renewables in the market. The model allows explaining the effects of competition and collaboration, by interpreting the cross imitation coefficients. Concerning the double-competition phase, we may notice for renewables the nonsignificant value of parameter p_{3d}, not reported in the results, reflecting the initial difficulty of these energy sources in entering the market and the consequent need for incentives to stimulate their initial growth. After the innovation phase, renewables have been experiencing strong internal growth, as testified by the high value of the within imitation coefficient, $q_{3d} = 0.58$. Also, the within imitation coefficient for nuclear is positive, $q_{2d} = 0.26$, which appears coherent with the increasing trend exhibited before the decline phase. On the other hand, the within imitation coefficient for coal is negative, $q_{1d} + \zeta = -0.28$, a finding that may be explained by the evident decreasing pattern of the series. Regarding the interplay between the three energy sources, the cross imitation coefficients help interpret how these energy sources have been interacting. Coal is characterized by a positive cross imitation effect, $q_{1d} = 0.68$, which suggests that nuclear and renewables collaborate with coal, which is however affected by intense negative internal dynamics. The nuclear cross imitation parameter is negative, $q_{2d} - \rho = -0.09$, indicating the competitive power exerted by coal and renewables that evidently affect the final part of the nuclear series. In contrast, renewables are characterized by a negative, but weak, cross imitation parameter, $q_{3d} - \xi = -0.001$, which shows that coal and nuclear compete with the new entrant energy sources, trying to limit their expansion. However, it should be observed that strong and rapid growth characterizes renewables, as already noticed from the value of the within imitation parameter. To aid interpretation, all these results are summarized in Tables 8.27 and 8.28.

Table 8.27   UCTT. Within imitation effects.

Coal	$q_{1d} + \zeta$	−0.28
Nuclear	q_{2d}	0.26
Renewables	q_{3d}	0.58

Table 8.28   UCTT. Cross imitation effects.

Coal	q_{1d}	0.68
Nuclear	$q_{2d} - \rho$	−0.09
Renewables	$q_{3d} - \xi$	−0.001

8.6.7 Recap

This case study has focused on the German energy transition, proposing some univariate and multivariate modeling for the energy sources employed for electricity generation. For each source, the best-fitting model has been selected and compared with the performance of the BM, trying to provide a clear interpretation of the involved parameters. The possibility to give a clear and straightforward meaning to the parameters and connect them with historical events is one of the valuable merits of diffusion models, which are not only used for predictive purposes but also to explain the reasons for specific behavior of data. The German energy market has proven to be a very dynamic environment, where the role of each energy source is rapidly changing as a result of well-planned policies, governmental commitment, and citizens' belief in sustainability and the transition to green technologies. This dynamism has been efficiently captured through competition modeling, by employing the UCRCD and the UCTT. The UCRCD has confirmed the results of Guidolin and Guseo (2016), according to which renewables have had a clear competitive impact on nuclear phaseout. The use of the UCTT to account for the additional effect of coal has led to notable results, given the difficulty of dealing with a system of three differential equations. Although the model does not provide an estimate of the effect that one energy source exerts on the other two, it still allows a broader understanding of the market and the complex interactions characterizing its evolution.

8.7 Growth of Video Conferencing

Innovative technologies and products may experience sudden success because of favorable market conditions, related to some social, economic, or environmental factor. This is the case of video conferencing, a technology that has been existing in markets for many years but that has experienced unprecedented growth during the Covid-19 pandemic, where populations were forced to stay at home for many weeks to respect social distancing and avoid the spread of the virus. This substantially modified the lifestyle and professional activities of millions of people, who had to continue to work, follow lectures, go to school, and communicate with others through an online platform. One of the most used video conferencing platforms is Zoom, developed by an American communications technology company headquartered in California and launched in 2013. In this case study, the rapid success of Zoom is studied by considering two different aspects, the temporal dynamics of its share price in the Nasdaq Stock Market, and the interest in Zoom according to searches in Google for the word "Zoom." These two processes show that the Covid-19 pandemic and the subsequent diffusion of the first vaccine against Covid-19 had a central role in the history of this technology.

8.7.1 Share Price

To have an idea of the rapid success of Zoom during the first months of the Covid-19 pandemic, it is interesting to analyze the price of Zoom shares traded on the Nasdaq Stock Market from the beginning of the year 2020 until the end of 2022, available at https://finance.yahoo.com. The share price can be considered a direct measure of the success of the technology and it is therefore useful to trace its dynamics through time. Figure 8.50 shows the weekly share price, and it is easy to notice the substantially bell-shaped behavior of the data, which would justify the application of innovation diffusion models. Also, one can observe that the growth in share price has been quite rapid, reaching a maximum of 500.11 U.S. dollars at the end of the year 2020, followed by a slow but unrelenting decline that started in December 2020. Trying to describe these data, the application of a simple BM produces a reasonable fit, as shown in Table 8.29 and Figure 8.51, well capturing the early phases of growth and the final decline of prices. What the BM cannot describe in a satisfactory way is the central part of the data, specifically the attainment of the maximum peak and the subsequent rapid fall at the end of 2020. A better model for this series is a GBM with one negative exponential shock, whose results are illustrated in Table 8.30 and Figure 8.52.

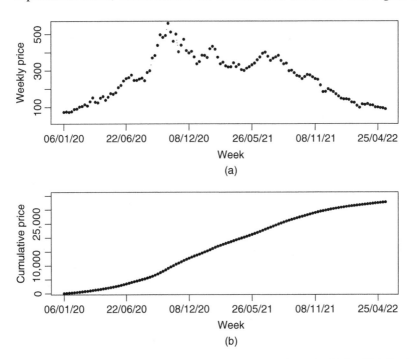

Figure 8.50 Zoom shares. Monthly price (a) and cumulative price (b) (in US dollars).

Table 8.29 Parameter estimates of BM for Zoom share price.

Parameter	Estimate	s.e.	Lower c.i.	Upper c.i.
m	34,407.5	192.1	34,030.9	34,784.1
p	0.0029	0.0001	0.0027	0.0030
q	0.04	0.0007	0.04	0.04

$R^2 = 0.9995$.

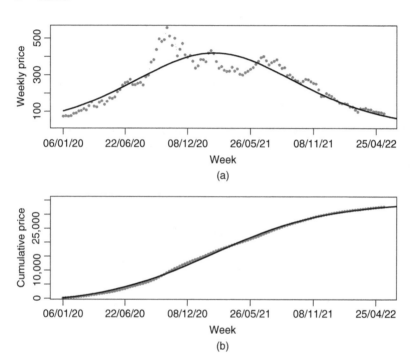

Figure 8.51 BM for Zoom shares. Weekly price (a) and cumulative price (b).

Overall, the model achieves a significantly higher level of fit with respect to the BM, and the comparison of the two models in terms of goodness of fit, $\tilde{R}^2 = 0.6$, confirms the selection of the GBM. The exponential shock is estimated to start in $a_1 = 49.7$, which is in the week starting on December 8, 2020, with a negative intensity, $c_1 = -0.43$ and negative memory, $b_1 = 0.02$ and may be reasonably imputed to the negative impact on share prices of Zoom of the exit in the market of the first Covid-19 vaccine, occurred at the beginning of December 2020, which obviously envisaged a solution to the pandemic. From a modeling point of view, the inclusion of the shock has allowed us to better describe the behavior of the

Table 8.30 Parameter estimates of GBM with one exponential shock for Zoom share prices.

Parameter	Estimate	s.e.	Lower c.i.	Upper c.i.
m	34,212.9	104.7	34,007.7	34,418.1
p	0.0018	0.00003	0.0018	0.0019
q	0.061	0.001	0.059	0.062
a_1	49.73	0.35	49.02	50.43
b_1	−0.02	0.002	−0.02	−0.01
c_1	−0.43	0.014	−0.46	−0.40

$R^2 = 0.9998$.

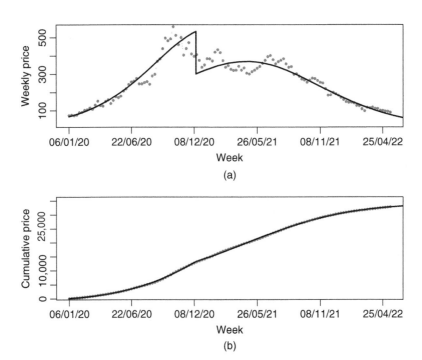

Figure 8.52 GBM with one exponential shock for Zoom shares. Weekly price (a) and cumulative price (b).

data, providing a meaningful interpretation of the negative shock. However, the model comparison shown in Figure 8.53 indicates that the predicted trajectories according to the BM and the GBM are very similar in the last portion of data, and thus suggests that the BM can be still an acceptable model for this series

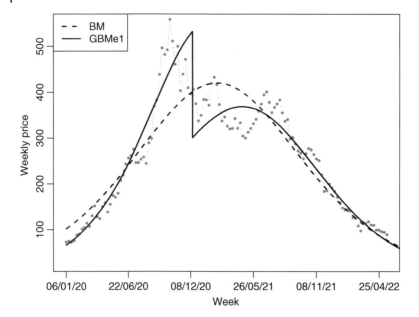

Figure 8.53 Model comparison. BM and GBM with one exponential shock for the weekly price of Zoom shares.

Table 8.31 SARMAX refinement for Zoom share price.

Parameter	ar1	ma1	sar1	sma1	intercept	λ
Estimate	0.9158	0.7416	0.7598	−0.9996	22.8968	0.9989
s.e.	0.0338	0.0606	0.2063	0.6756	76.6963	0.0039

$AIC = 1187.61$.

for prediction purposes. After model selection, a usual SARMAX refinement efficiently completes the analysis, by describing the residual variability due to seasonality and autocorrelation. The results of this procedure are shown in Table 8.31 and Figure 8.54.

8.7.2 Google Searches

Another interesting indicator of the success of Zoom is the weekly series of Google searches in the world for "Zoom" starting from October 2019, according to the data retrieved from Google Trends (https://trends.google.it/trends/). Note that the word "Zoom" may include other objects (such as zooms for cameras) but these searches represent a negligible portion. By inspecting the data in Figure 8.55, one

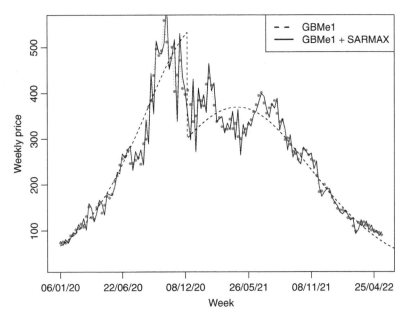

Figure 8.54 SARMAX refinement after GBM with one exponential shock for weekly price of Zoom shares.

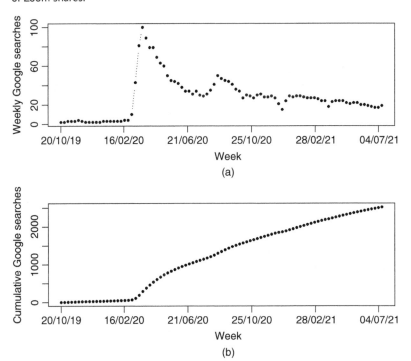

Figure 8.55 Google searches of "Zoom." Weekly searches (a) and cumulative searches (b).

may see that the sudden interest in Zoom arising in March 2020 was very strong, whereas before 2020, it was nearly absent. It is also evident that such growth in interest in Zoom was very rapid and transient, because, after a peak reached at the end of March 2020, then the data started to decrease quite quickly. This phenomenon has a straightforward explanation: once people installed and started to use Zoom or other video conferencing platforms, they did not need to search for information about these technologies on the web anymore. Looking at the behavior of data, it is clear that a simple BM would not be able to describe the exponential increase in searches registered in March 2020, thus the modeling strategy suggests directly fitting a GBM with one exponential shock, whose fit is especially good, as may be seen in Figure 8.56.

Table 8.32 reports parameter estimates of the model and shows that the explosion of interest started in the first week of March 2020, $a_1 = 21.11$, with an extremely strong intensity, $c_1 = 11.44$ but also with a strongly decaying effect, $b_1 = -0.25$. The values of the diffusion parameters, $p = 0.0007$ and $q = 0.04$ highlight a very slow growth process in its early stages, as indicated by the weak innovative component. This underlines the central role played by the exponential

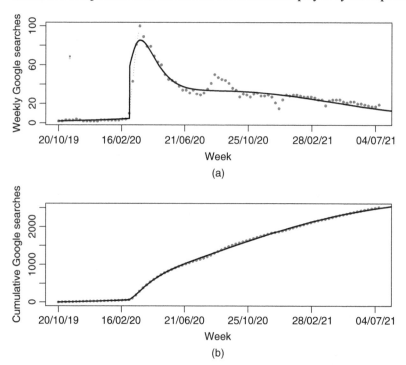

(a)

(b)

Figure 8.56 GBM with one exponential shock for Google searches of "Zoom." Weekly searches (a) and cumulative searches (b).

Table 8.32 Parameter estimates of GBM with one exponential shock for Google searches of "Zoom".

Parameter	Estimate	s.e.	Lower c.i.	Upper c.i.
m	2863.9	25.9	2813.2	2914.6
p	0.0007	0.0001	0.0005	0.0008
q	0.04	0.0009	0.04	0.05
a_1	21.11	0.2323	20.65	21.56
b_1	−0.25	0.0131	−0.28	−0.22
c_1	11.44	0.7132	10.04	12.83

$R^2 = 0.9998$.

Table 8.33 SARMAX refinement for Google searches of "Zoom.".

Parameter	ar1	ma1	sma1	intercept	λ
Estimate	0.9155	0.5564	−0.1156	−0.1915	1.0019
s.e.	0.0402	0.0782	0.2036	10.9422	0.0071

$AIC = 557.16$.

shock in explaining the increase in interest in Zoom, and therefore the strong boost to the adoption of Zoom is implied by the Covid-19 pandemic. A SARMAX refinement efficiently completes the analysis, as shown in Table 8.33 and Figure 8.57.

8.7.3 Recap

The growth of a new technology may be measured by considering its growth in the financial market and its success on the web. In this case study, the rapid market expansion of video conferencing Zoom has been analyzed, by modeling the price of company shares traded in the financial markets and the online searches for the word "Zoom" according to Google Trends. The two observed processes clearly show the role that the Covid-19 pandemic has had on the success of this technology and, from a modeling perspective, highlight the crucial role of the GBM in capturing the intense perturbations characterizing the data.

8.8 Diffusion of a Scientific Paper

So far, it has been shown that innovation diffusion models are generally used to model the growth of a new product or technology in a market. However, with the

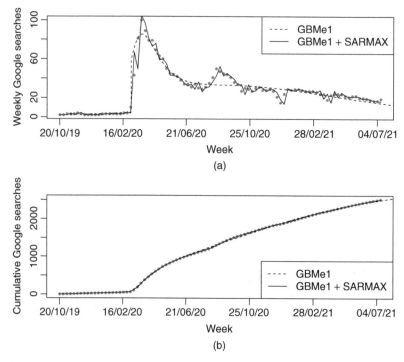

Figure 8.57 SARMAX refinement after GBM with one exponential shock for Google searches of "Zoom." (a) Weekly searches, (b) cumulative searches.

term innovation, we also indicate new ideas or new ways of doing things, whose spread in society may show an evolution that is coherent with innovation diffusion theory. To provide a different example dealing with the diffusion of an idea, here we consider the success of a scientific paper in the academic community by analyzing the number of citations it has received from its publication on. In fact, the number of citations may be seen as a direct measure of the success and importance of a particular research topic among scholars and, to some extent, of the importance of the researchers working on it. To describe a paradigmatic case, it is here considered the number of citations according to the freely available information provided in Google Scholar (https://scholar.google.com) of the article "A New Product Growth for Model Consumer Durables" by Frank Bass, published in Management Science in 1969, where the BM was proposed for the first time. After the publication of the BM, the literature on innovation diffusion in the marketing field started to expand and the interest in the prediction of sales of new products stimulated a new research field to which also this book belongs, dealing with quantitative marketing and strategic decisions concerning product takeoff, market saturation, and life cycle determinants. Figure 8.58 shows the time

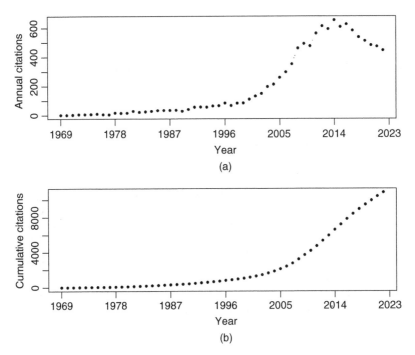

Figure 8.58 Citations of the paper by Bass (1969). Annual data (a) and cumulative data (b).

series of yearly citations of the paper from 1969 to 2022, whose growth started to be evident in the mid-90s, whereas the first part of the process was characterized by a flat trend. This slow increase in citations observed during the 70s and 80s may be justified considering two aspects: on the one hand, until the 90s there was no Internet so the academic research activity was very different, with much more difficulty in retrieving the literature on a particular topic; on the other hand, in 1994 Bass, Krishnan, and Jain published the paper "Why the Bass Model Fits without Decision Variables" in Marketing Science, proposing the Generalized Bass model, which, as also discussed in this book, provided a larger perspective on life cycle modeling by accounting for the effect of marketing mix variables. Also, the first reviews on innovation diffusion modeling were published in the 90s, e.g. Mahajan et al. (1990) and Bass (1995), contributing to the spread of knowledge on the topic and the celebrity of Bass (1969). The observed data also show that the growth of citations has apparently reached a peak in 2014, after which a slow decline started. This may be interpreted as a sign that the BM is becoming a less central model in the scientific community that studies innovation diffusion processes. Indeed, the different examples provided in this book have shown that the BM is often

Table 8.34 Parameter estimates of BM for annual citations of the paper by Bass (1969).

Parameter	Estimate	s.e.	Lower c.i.	Upper c.i.
m	15,107.5	438.6	14,247.8	15,967.2
p	0.0001	0.00001	0.0001	0.0001
q	0.16	0.0037	0.15	0.17

$R^2 = 0.9990$.

an insufficient model to describe life cycle dynamics, which exhibit shapes better described through more complex models, such as the GBM and GGM. In trying to model these data, the application of a BM is a straightforward choice, whose results are outlined in Table 8.34 and Figure 8.59.

By inspecting the model fit in Figure 8.59, the BM appears to be a fair solution for the data, being able to capture the mean trajectory in an acceptable way. However, the prediction according to the BM tends to overestimate the portion of data between 1996 and 2007, so other modeling alternatives are evaluated. Regarding parameter estimates, it is interesting to observe that the innovative component

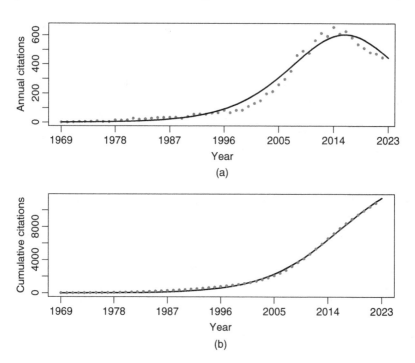

Figure 8.59 BM for citations of the paper by Bass (1969). Annual data (a) and cumulative data (b).

Table 8.35 Parameter estimates of GBM with one rectangular shock for yearly citations of Bass (1969).

Parameter	Estimate	s.e.	Lower c.i.	Upper c.i.
m	13,391.1	71.1	13,251.9	13,530.2
p	0.0007	0.0002	0.0004	0.0010
q	0.18	0.0013	0.18	0.18
a_1	−1.64	1.9651	−5.49	2.20
b_1	36.26	0.1494	35.97	36.56
c_1	−0.41	0.0120	−0.44	−0.39

$R^2 = 0.9999$.

has been very weak, $p = 0.0001$, which explains the slow takeoff of the citations, while the imitative component has been quite strong, $q = 0.16$, denoting intense word-of-mouth and learning behavior of researchers. In order to improve the fit, among various modeling options, the best selected is a GBM with one rectangular shock. The satisfactory fit obtained may be appreciated in Table 8.35 and Figure 8.60. The graphical display of the results shows that the shock has been

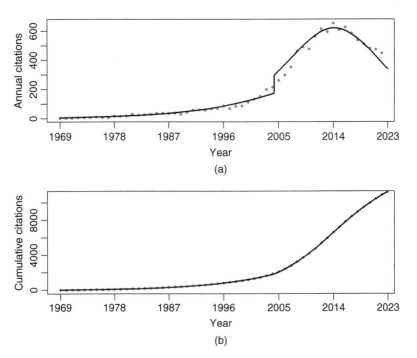

Figure 8.60 GBM with one rectangular shock for citations of Bass (1969). Annual data (a) and cumulative data (b).

estimated to start at the beginning of the series and last until the year 2004. Parameter estimates confirm the significant presence of a negative shock, $c_1 = -0.41$, starting in 1969, $a_1 \simeq 0$, and ending in 2004, $b_1 = 36.2$. This negative perturbation may be interpreted as an inertial effect or a negative externality, preventing the diffusion of the paper, which lasted until the beginning of the 2000s. Again, this negative externality may be due to the absence of the Internet in the first years, which surely prevented the widespread awareness of the relevant literature. The improvement obtained with respect to the simple BM is evident, since $\tilde{R}^2 = 0.90$.

The comparison between the two models in terms of goodness of fit is proposed in Figure 8.61, where the better performance of the GBM is more visible.

8.8.1 Recap

This case study has focused on the growth dynamics of a scientific paper, measured in terms of the citations it has received from its publication on. The example is useful to show that also the spread of new ideas, contained for example in a scientific paper or a book, follows a trajectory that can be modeled with the innovation diffusion approach. This application highlights the role of the GBM in capturing the slow takeoff of citations in the first years since publication and the sharp effect played by exogenous variables, such as Internet diffusion and publication of review

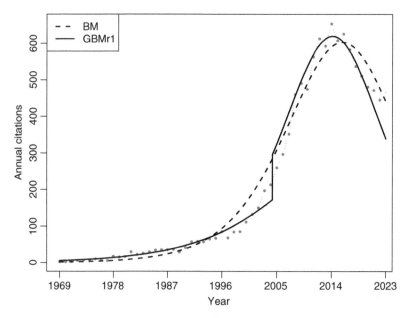

Figure 8.61 Model comparison. BM and GBM with one rectangular shock for annual citations of Bass (1969).

papers. The example is also interesting in showing that a shock may be estimated at the beginning of the diffusion process and does not need to be necessarily in the central part of the observed data.

8.9   Diffusion of Internet Usage

This case study takes a different perspective with respect to the others analyzed in this chapter, by considering a large number of time series, for which a common modeling solution is proposed. The phenomenon analyzed is the diffusion of the Internet in a large number of countries around the world and, through the modeling, we will try to understand whether there are countries that share a similar diffusion pattern. The literature on diffusion modeling has widely discussed the difference between countries in adopting innovations, showing different diffusion patterns that may be the result of many factors, such as a country's economic and social conditions, the availability of the necessary infrastructures, and a market that is ready and prepared to receive the novelty or not. Before entering into the details of the data here analyzed, we may make the hypothesis that such differences between countries will shape different historical trajectories for Internet adoption.

The Internet's history goes back some decades by now, but it can be still considered a technology at the beginning of its life cycle because its characteristics and usage are rapidly evolving over time. The invention of the World Wide Web dates back to 1989, thanks to the work of Tim Berners-Lee, who created a system to share information through a network of computers. The parallel development of technologies such as fiber optic cables in the mid of the 90s allowed a dramatic spread of the Internet, with the advent of communication technologies such as emails, instant messaging, telephone calls, and video chats. The Internet has had a possibly stronger impact on society, transforming cultural and social habits, economic rules, and lifestyles with the rise of social networks in the 2000s, and the advent of the so-called Web 2.0. In this sense, the adoption of social networks and their widespread usage has been stimulated by the parallel diffusion of smartphones allowing a substantial change in communication between people. The Internet is continuing to grow and evolve, driven by increasing amounts of online information, commerce, entertainment, and communication services.

The diffusion of the Internet started in the 1990s, at least in some regions of the world. By the year 2000, almost half of the population in the United States was accessing information online. However, in many other regions of the world, the Internet has not yet had much impact: as reported by Roser et al. (2018), 93% in the East Asia and Pacific region and 99% in South Asia and in Sub-Saharan Africa were still offline in 2000. According to Roser et al. (2018), in 2016, three-quarters of people in the United States were online and during these years many other countries grew significantly: in Malaysia, 79% used the Internet, in Spain and Singapore 81%, in France 86%, in South Korea and Japan 93%, in Denmark and Norway 97%, and Iceland tops the ranking with 98% of the population online. At the other end of the story, there are still countries where almost nothing has changed since 1990. In the very poorest countries of Africa, less than 5% are online.

The data here considered, available at https://ourworldindata.org, refer to 224 countries for which the diffusion of the Internet has been observed starting from 1990 to 2020. Figure 8.62 displays the case of some of these countries, namely Myanmar, Portugal, Thailand, the United Kingdom, and the United States that have been selected as paradigmatic examples of the regional difference in diffusion patterns. Note that the data have been here normalized between 0 and 1 in order to show together countries whose diffusion is very different in terms of size (e.g. Internet usage in Myanmar would disappear compared to that of the United States if one used the absolute data). By observing the normalized data, one can note that the United States, the United Kingdom, and Portugal started quite early, and the diffusion process has a clearly increasing behavior in the mid-90s, showing some evidence of saturation between 2010 and 2020. Thailand had a slower diffusion

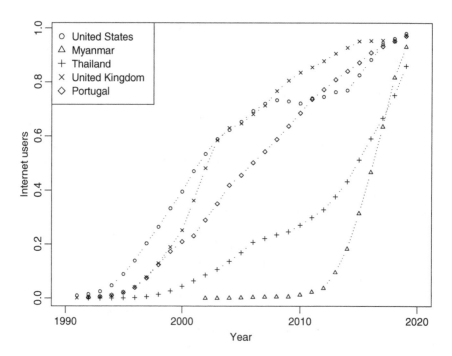

Figure 8.62 Percentage Internet diffusion in some countries in the world.

until the first years of the 2000s and registered an acceleration by then, with ongoing growth. The last case is that of Myanmar, for which the history of Internet usage is more recent, starting in 2002, and after low growth in the early stages, is registering an explosive expansion in the last decade. These examples suggest that some countries are already in an advanced phase of the Internet life cycle, while others have begun ten years later and are still increasing their user base.

Although every country has its own history and peculiar traits that would deserve a specific focus, the analysis proposed in this case study is based on the application of a unique model to all the 224 series, in order to capture some possible regularities occurring between them. The model selected to fit all the different trajectories in an acceptable way is the BM, producing some interesting findings based on the estimated parameters. The BM has been estimated on the original data, i.e. not normalized. The BM has been selected as the most suitable model because the purpose of the analysis is not a specific prediction for each country, but the description of some growth profiles, which is guaranteed by the BM, for its simplicity, identifiability, and the easiness of interpretation of the involved parameters. To comment on the model findings, rather than studying parameter estimates for each country, it is more convenient to visualize how the parameters are distributed across regions, as may be inspected through some world maps, illustrated in Figures 8.63–8.65. This helps identify some meaningful groups.

Once the distribution of each parameter has been analyzed, it is useful to investigate the possible relationship between pairs of parameters, namely m vs p, m vs q, and p vs q. These relationships are displayed in Figures 8.66–8.68 and offer a complementary view of similarities and differences among countries. In the three plots, all the analyzed countries are displayed, but for reasons of clarity, just some of these have been labeled. In Figure 8.66, the association between the estimated market potential m and the innovation coefficient p is illustrated. It may be easily

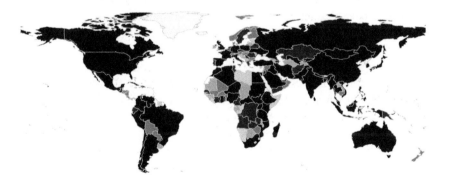

Figure 8.63 Groups of countries according to the market potential m.

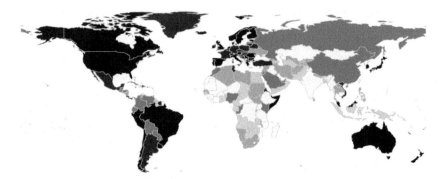

Figure 8.64 Groups of countries according to the innovation coefficient p.

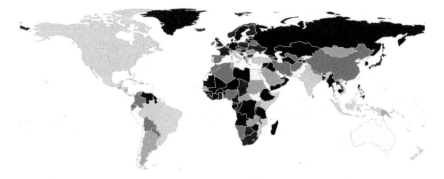

Figure 8.65 Groups of countries according to the imitation coefficient q.

observed that most countries have a quite small p coefficient, suggesting a general initial difficulty in adopting the new technology in the early stages of its life. The only exceptions are the United States, Canada, Australia, and a few others, which have been characterized by a high level of innovativeness and openness toward new technology, being the pioneers in Internet usage. On the other hand, China and India, which have the biggest market potential, present particularly low innovation coefficients, below 0.025, suggesting that these two countries have been laggards in Internet usage. This is not a surprising fact, since many countries like Brazil, India, China, and Russia started to grow significantly at the beginning of the 2000s. The distribution of parameter p with respect to the market size m may be divided into three regions, the first with $p < 0.0025$, including the majority of countries, the second with $0.0025 < p < 0.01$, where several European countries appear, such as Finland, Italy, Netherlands, Norway, Portugal, Sweden, and the United Kingdom, and the third with $p > 0.01$, including the United States, Canada, and Australia.

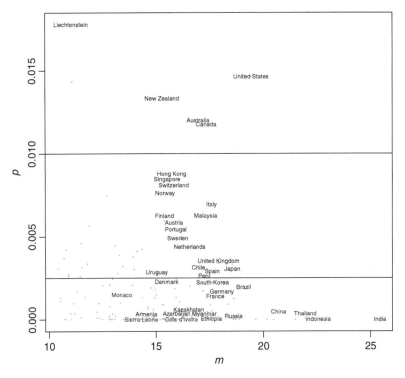

Figure 8.66 Relationship between *m* and *p* in each group.

A different situation may be observed in the distribution of parameter q with respect to m, displayed in Figure 8.67. As visible, the points are more scattered, and it seems reasonable to divide them into two regions, fixing a threshold at $q = 0.4$. Among the countries characterized by a high imitation coefficient, $q > 0.4$, it is interesting to observe that the majority are African, such as Sierra Leone, Ethiopia, the Republic of Congo, and Cote d'Ivoire, or Post-Soviet states, like Armenia, Azerbaijan, and Kazakhstan. In both cases, these are states that started to develop recently and are showing a dramatic increase in their access to information and communication technologies. On the other hand, among those with $q < 0.4$, the United States exhibit a particularly low imitation coefficient.

Finally, Figure 8.68 describes the relationship between coefficients p and q and indicates an essentially negative association between the innovation and the imitation components. Countries with a high innovation coefficient, i.e. $p > 0.01$, such as the United States, Canada, and Australia, are characterized by a relatively low imitation coefficient, $q \simeq 0.2$, whereas those with $0.025 < p < 0.01$ typically exhibit an average value of imitation coefficient, i.e. $0.2 < q < 0.4$. This group includes several northern European states, which are well known for being leaders

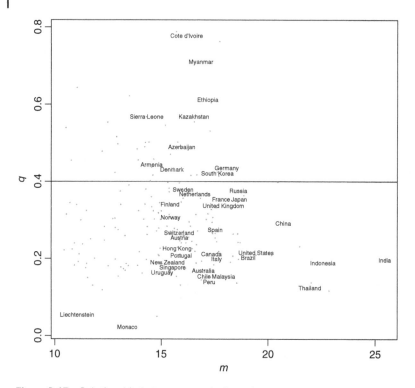

Figure 8.67 Relationship between *m* and *q* in each group.

in information and communication technologies, and some East Asian nations, like Hong Kong and Singapore, which have been historically termed the "Asian tigers," together with South Korea and Taiwan, for their rapid industrialization and exceptionally high growth rates between the early 60s and 90s.

Concerning the majority of countries, which show an innovation coefficient below 0.0025, we may observe that most of them also have an imitation coefficient below the threshold of 0.4, whereas a smaller portion of nations is characterized by a high imitation coefficient, $q > 0.4$, most of which are developing countries.

Indeed, the analysis of the relationship between innovation and imitation coefficients provides more interesting results and allows the detection of four groups of countries, which may be best appreciated in Figure 8.69. Black countries are those with high innovation and low imitation, $p > 0.1$ and $q < 0.4$, namely the United States, Canada, Australia, and New Zealand. Dark gray countries are those with average innovation and low imitation, $0.0025 < p < 0.1$ and $q < 0.4$, essentially identifying a cluster of European countries, namely Austria, Italy, Portugal, Spain, Switzerland, the United Kingdom, and Nordic countries. Out of Europe, this group includes developed East Asian nations, namely Japan, Indonesia, and

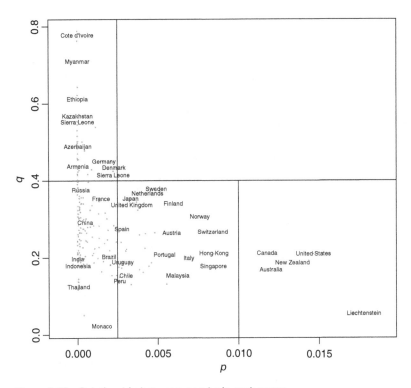

Figure 8.68 Relationship between *p* and *q* in each group.

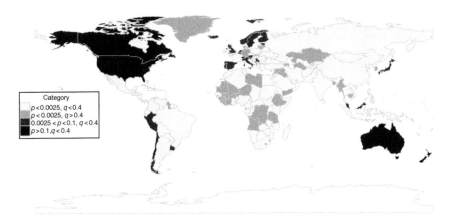

Figure 8.69 Groups according to *p* and *q*.

Singapore. This group also comprises three South-American countries, i.e. Chile, Perù, and Uruguay. The group of countries colored in medium gray is composed of those with low innovation coefficient, $p < 0.0025$, but a high imitation, $q > 0.4$, a pattern that characterizes several African and some Post-Soviet states, which are experiencing dramatic growth in these years. In Europe, this combination of parameters is estimated in Germany and Denmark, denoting a notable exception with respect to typical European behavior, as commented earlier. Finally, the rest of the countries, colored in light gray, are those with $p < 0.0025$ and $q < 0.4$, representing the largest group. From a general perspective, we may take as peculiar situations those groups presenting either a high level of innovation, $p > 0.1$, or a high level of imitation, $q > 0.4$, whereas the majority of countries are characterized by $0 < p < 0.1$ and $q < 0.4$. To interpret these results, we may consider the countries with $p > 0.1$ as the "innovators," i.e. pioneers in the adoption of the technology, the largest group of countries with $0 < p < 0.1$ and $q < 0.4$ as the "majority," which may be divided into two sub-groups, the "early adopters," i.e. those countries with $0.0025 < p < 0.1$ and the "late adopters", i.e. those with $p < 0.0025$.

8.9.1 Recap

This case study has taken a different perspective with respect to the previous ones, by considering a large number of time series reporting the adoption of the Internet at the country level. Despite the great variety of diffusion patterns in the data, the BM has been chosen as a unique modeling tool for the phenomenon. The availability of estimated parameters m, p, and q for hundreds of countries has suggested using this information to identify some groups of countries that show similar diffusion patterns. The results obtained through this clustering procedure based on the BM parameters have led to very interesting and meaningful conclusions, showing that the BM, thanks to its simplicity, can be employed in the case of large datasets to extract well-identified profiles of adoption by the means of three interpretable parameters.

References

Frank M Bass. A new product growth model for consumer durables. *Management Science*, 15(5):215–227, 1969.

Frank M Bass. Empirical generalizations and marketing science: A personal view. *Marketing Science*, 14(3 supplement):G6–G19, 1995.

Mariangela Guidolin and Renato Guseo. Technological change in the US music industry: Within-product, cross-product and churn effects between competing blockbusters. *Technological Forecasting and Social Change*, 99:35–46, 2015.

Mariangela Guidolin and Renato Guseo. The German energy transition: Modeling competition and substitution between nuclear power and renewable energy technologies. *Renewable and Sustainable Energy Reviews*, 60:1498–1504, 2016.

Mariangela Guidolin and Piero Manfredi. Innovation diffusion processes: Concepts, models, and predictions. *Annual Review of Statistics and Its Application*, 10(1):451–473, 2023. doi: 10.1146/annurev-statistics-040220-091526.

Vijay Mahajan, Eitan Muller, and Frank M Bass. New product diffusion models in marketing: A review and directions for research. *Journal of Marketing*, 54(1):1–26, 1990.

Max Roser, Hannah Ritchie, and Esteban Ortiz-Ospina. The internet's history has just begun. *Our World in Data*, 2018. URL https://ourworldindata.org/internet-history-just-begun.

9

Modeling a Diffusion Process

9.1 Statistical Modeling

In 1976, the statistician George Box published an article entitled "Science and Statistics" in the Journal of the American Statistical Association (Box, 1976), where he discussed aspects of the scientific method and statistical modeling. The key principles for a model building he focused on were a continuous iteration between theory and practice, producing a feedback loop where the scientist must have the flexibility to recognize the errors—particularly his own—and avoid the risk "to fall in love with his model." Because of the inherent presence of errors in any statistical model, Box stated that all models are essentially wrong, but clarified that excessive elaboration is not the right direction to obtain a "correct" model. On the contrary, one should try to describe phenomena with simple and evocative structures, complying with the criterion of model parsimony. Box also added the idea of being selectively concerned about the errors in a model, writing that "since all models are wrong the scientist must be alert to what is *importantly* wrong." Some years later, he provided a more general statement about statistical modeling, where he clarified that the only question of interest is whether a model is "illuminating and useful" (Box, 1979).

Following the definition provided by Azzalini and Scarpa (2012), we may define a model as "a simplified representation of the phenomenon of interest, functional for a specific objective": it is, therefore, necessary to reduce aspects that are not essential for the analysis while maintaining those functional to the objective. This will lead to the conclusion that "inevitably, the model will be wrong-but it must be wrong to be useful" (Azzalini and Scarpa, 2012).

In this book, for example, the Guseo–Guidolin model (GGM) has been developed by making some reasonable hypotheses on how communication dynamics may influence the creation of a market potential and therefore the final diffusion process. In the Generalized Bass Model (GBM), the structured shocks, both exponential and rectangular, have been introduced to capture the effect of

Innovation Diffusion Models: Theory and Practice, First Edition. Mariangela Guidolin.
© 2024 John Wiley & Sons Ltd. Published 2024 by John Wiley & Sons Ltd.
Companion Website: www.wiley.com/go/innovationdiffusionmodels/

exogenous perturbations that may alter the growth in a significant way. These are of course simplified representations of reality, and therefore "wrong" models, but they try to capture the most relevant pattern in a diffusion process.

9.2 To Explain or to Predict

The debate on what statistical modeling is and should be has always been rich and multifaceted, but become particularly intense in the 2000s, starting from Leo Breiman's paper "Statistical Modeling: The Two Cultures" (Breiman, 2001), where he proposed the idea that there are two cultures in the use of statistical modeling: one assumes that the data are generated by a given stochastic data model, while the other uses algorithmic models and treats the data mechanism as unknown. In Breiman's view, the statistical community has been always focused on the first culture, which has sometimes led to questionable conclusions, while algorithmic modeling could be used as a more flexible approach to solve problems and extract information from data. In commenting on Breiman's paper, Sir David Cox observed that "formal models are useful and often almost, if not quite, essential for incisive thinking. Descriptively appealing and transparent methods with a firm model base are ideal. Notions of significance tests, confidence intervals, posterior intervals, and all the formal apparatus of inference are valuable tools to be used as guides, *but not in a mechanical way*" (Cox, 2001). Therefore, he warned about automatic model selection, because "automatic methods of model selection (and of variable selection in regression-like problems) are to be shunned or, if use is absolutely unavoidable, are to be examined carefully for their effect on the final conclusions. Unfocused tests of model adequacy are rarely helpful" (Cox, 2001).

These apparently distant perspectives on statistical modeling have found an opportunity for reconciliation in the article by Galit Shmueli "To Explain or to Predict" (Shmueli, 2010), where statistical modeling has been defined depending on its functional purpose, i.e. if the model is used for explanatory or predictive goals. Shmueli defined explanatory modeling as "the use of statistical models for testing causal explanations," whereas predictive modeling is "the process of applying a statistical model or data mining algorithm to data for the purpose of predicting new or future observations." Being able to recognize the difference between explanatory and predictive modeling is critical for the practical implications that this difference can have on each step of the modeling process.

9.2.1 To Explain

In trying to learn some important lessons from this debate on statistical modeling, one could apply previous ideas and define innovation diffusion models as a parsimonious approach, built on ordinary differential equations, which provides a simplified and useful representation of a diffusion process over time, exploiting a continuous exchange between theory and practice. All the examples discussed in the book have shown that model building and model selection cannot be performed on automatic procedures, but always require a careful evaluation from the researcher, who needs to confront the practical conclusions of a model choice and always be aware of the assumptions made when applying a certain class of models. Just like the Bass Model (BM), also its generalizations illustrated throughout the chapters, like the GBM (Chapter 3), the GGM (Chapter 4), and the UCRCD model (Chapter 6), have been developed to be simple, yet powerful tools for explaining diffusion phenomena, that depart from the classical shape described by the Bass Model, and account for the effect of structured shocks, complex communication dynamics, and competition. From this perspective, diffusion models have a clear explanatory goal, because they try to explain the reasons *why* a given diffusion process has a certain shape. For example, in Section 8.6.4, the diffusion of renewables in Germany has been studied with the BM and the GBM with one exponential shock and it has been clear that the inclusion of the shock has been essential for understanding the role played by incentives in stimulating the market for this kind of energy. The use of the GGM has been crucial in many examples, such as in Sections 8.4 and 8.5, for appreciating the role of a dynamic market potential and how the creation of a group of informed consumers can play a decisive role for a product's success.

9.2.2 To Predict

Innovation diffusion models can also be used for predictive purposes, and tell us *what* will be the shape of a given diffusion process. As we have seen in the first case study in Chapter 8, Section 8.2, out-of-sample predictions may be easily produced and employed for forecasting the future evolution of a product's life cycle. However, one should bear in mind that such forecasting exercise needs to be performed with care since different models may produce very diversified scenarios, as shown in Section 8.2. Also, the amount of available data and the presence or not of crucial data points, like the maximum peak, should always be accounted for, because of the strong implications they can have on the possibility to handle reliable predictions, as discussed in the second case study of Chapter 8, Section 8.3. Finally, the last case study proposed in Section 8.9, concerning the use of a large number of time series, has suggested a different way of employing diffusion models from a predictive point of view: similar patterns can be described according to a common model and the information extracted can serve as a baseline to construct some well-identified diffusion profiles, which may be used for "guessing by analogy" approaches where little or no data are available.

9.3 Conclusion

The theory and practice of innovation diffusion models described in this book have taught us that building a model, selecting among different models, using models for explanation or prediction, and final evaluation always requires understanding and accurate consideration of the problem at hand by the analyst. All the practical cases proposed have shown that there are no ready-made solutions and automatic procedures that fit every situation are not available (and perhaps not desirable). Whatever the purpose of the analysis, be it explanatory or predictive, the type of data considered, and the model finally employed, there is always a need for handmade work, which highly depends on a researcher's experience, knowledge, observation, and imagination.

"Modelling in science remains, partly at least, an art" (McCullagh and Nelder, 1983).

References

Adelchi Azzalini and Bruno Scarpa. *Data Analysis and Data Mining: An Introduction.* OUP USA, 2012.

George EP Box. Science and statistics. *Journal of the American Statistical Association,* 71(356):791–799, 1976.

George EP Box. Robustness in the strategy of scientific model building. In *Robustness in Statistics,* pages 201–236. Elsevier, 1979.

Leo Breiman. Statistical modeling: The two cultures (with comments and a rejoinder by the author). *Statistical Science,* 16(3):199–231, 2001.

David R Cox. [statistical modeling: The two cultures]: Comment. *Statistical Science,* 16(3):216–218, 2001.

Peter McCullagh and John A Nelder. 1989. *Generalized Linear Models,* volume 37. CRC Press, 1983.

Galit Shmueli. To explain or to predict? *Statistical Science,* 25(3):289–310, 2010.

References

Guillermo Abramson and Damián H Zanette. Statistics of extinction and survival in Lotka-Volterra systems. *Physical Review E*, 57(4):4572, 1998.

Hirotsugu Akaike. A new look at the statistical model identification. *IEEE Transactions on Automatic Control*, 19(6):716–723, 1974.

Adelchi Azzalini and Bruno Scarpa. *Data Analysis and Data Mining: An Introduction.* OUP USA, 2012.

Frank M Bass. A new product growth model for consumer durables. *Management Science*, 15(5):215–227, 1969.

Frank M Bass. Empirical generalizations and marketing science: A personal view. *Marketing Science*, 14(3 supplement):G6–G19, 1995.

Frank M Bass, Trichy V Krishnan, and Dipak C Jain. Why the Bass model fits without decision variables. *Marketing Science*, 13(3):203–223, 1994.

Alessandro Bessi, Mariangela Guidolin, and Piero Manfredi. The role of gas on future perspectives of renewable energy diffusion: Bridging technology or lock-in? *Renewable and Sustainable Energy Reviews*, 152:111673, 2021.

George EP Box. Science and statistics. *Journal of the American Statistical Association*, 71(356):791–799, 1976.

George EP Box. Robustness in the strategy of scientific model building. In *Robustness in Statistics*, pages 201–236. Elsevier, 1979.

George EP Box, Gwilym M Jenkins, Gregory C Reinsel, and Greta M Ljung. *Time Series Analysis: Forecasting and Control.* New York: Wiley, 2015.

Leo Breiman. Statistical modeling: The two cultures (with comments and a rejoinder by the author). *Statistical Science*, 16(3):199–231, 2001.

Anita M Bunea, Pompeo Della Posta, Mariangela Guidolin, and Piero Manfredi. What do adoption patterns of solar panels observed so far tell about governments' incentive? Insights from diffusion models. *Technological Forecasting and Social Change*, 160:120240, 2020.

Innovation Diffusion Models: Theory and Practice, First Edition. Mariangela Guidolin.
© 2024 John Wiley & Sons Ltd. Published 2024 by John Wiley & Sons Ltd.
Companion Website: www.wiley.com/go/innovationdiffusionmodels/

Anita M Bunea, Mariangela Guidolin, Piero Manfredi, and Pompeo Della Posta. Diffusion of solar PV energy in the UK: A comparison of sectoral patterns. *Forecasting*, 4(2):456–476, 2022.

Deepa Chandrasekaran and Gerard J Tellis. A critical review of marketing research on diffusion of new products. *Review of Marketing Research*, 3:39–80, 2007.

Wesley M Cohen and Daniel A Levinthal. Absorptive capacity: A new perspective on learning and innovation. *Administrative Science Quarterly*, 35:128–152, 1990.

David R Cox. [statistical modeling: The two cultures]: Comment. *Statistical Science*, 16(3):216–218, 2001.

Alessandra Dalla Valle and Claudia Furlan. Forecasting accuracy of wind power technology diffusion models across countries. *International Journal of Forecasting*, 27(2):592–601, 2011.

George S Day. The product life cycle: Analysis and applications issues. *Journal of Marketing*, 45(4):60–67, 1981.

James Durbin and Geoffrey Watson. Testing for serial correlation in least squares regression: I. *Biometrika*, 37(3/4):409–428, 1950.

Jan Fagerberg, David C Mowery, Richard R Nelson, et al. *The Oxford Handbook of Innovation*. Oxford: Oxford University Press, 2005.

Claudia Furlan and Cinzia Mortarino. Forecasting the impact of renewable energies in competition with non-renewable sources. *Renewable and Sustainable Energy Reviews*, 81:1879–1886, 2018.

Claudia Furlan, Mariangela Guidolin, and Renato Guseo. Has the Fukushima accident influenced short-term consumption in the evolution of nuclear energy? An analysis of the world and seven leading countries. *Technological Forecasting and Social Change*, 107:37–49, 2016.

Claudia Furlan, Cinzia Mortarino, and Mohammad Salim Zahangir. Interaction among three substitute products: An extended innovation diffusion model. *Statistical Methods & Applications*, 30(1):269–293, 2021.

Jelle J Goeman and Aldo Solari. Multiple testing for exploratory research. *Statistical Science*, 26(4):584–597, 2011.

Mariangela Guidolin and Tansu Alpcan. Transition to sustainable energy generation in Australia: Interplay between coal, gas and renewables. *Renewable Energy*, 139:359–367, 2019.

Mariangela Guidolin and Renato Guseo. Modelling seasonality in innovation diffusion. *Technological Forecasting and Social Change*, 86:33–40, 2014.

Mariangela Guidolin and Renato Guseo. Technological change in the US music industry: Within-product, cross-product and churn effects between competing blockbusters. *Technological Forecasting and Social Change*, 99:35–46, 2015.

Mariangela Guidolin and Renato Guseo. The German energy transition: Modeling competition and substitution between nuclear power and renewable energy technologies. *Renewable and Sustainable Energy Reviews*, 60:1498–1504, 2016.

Mariangela Guidolin and Renato Guseo. Has the iPhone cannibalized the iPad? An asymmetric competition model. *Applied Stochastic Models in Business and Industry*, 36(3):465–476, 2020.

Mariangela Guidolin and Piero Manfredi. Innovation diffusion processes: Concepts, models, and predictions. *Annual Review of Statistics and Its Application*, 10(1):451–473, 2023. doi: 10.1146/annurev-statistics-040220-091526.

Mariangela Guidolin and Cinzia Mortarino. Cross-country diffusion of photovoltaic systems: Modelling choices and forecasts for national adoption patterns. *Technological Forecasting and Social Change*, 77(2):279–296, 2010.

Mariangela Guidolin, Renato Guseo, and Cinzia Mortarino. Regular and promotional sales in new product life cycles: Competition and forecasting. *Computers & Industrial Engineering*, 130:250–257, 2019.

Renato Guseo. Strategic interventions and competitive aspects in innovation life cycle. Technical report, Department of Statistical Sciences, University of Padua, 2004.

Renato Guseo and Alessandra Dalla Valle. Oil and gas depletion: Diffusion models and forecasting under strategic intervention. *Statistical Methods and Applications*, 14(3):375–387, 2005.

Renato Guseo and Mariangela Guidolin. Modelling a dynamic market potential: A class of automata networks for diffusion of innovations. *Technological Forecasting and Social Change*, 76(6):806–820, 2009.

Renato Guseo and Mariangela Guidolin. Cellular automata with network incubation in information technology diffusion. *Physica A: Statistical Mechanics and its Applications*, 389(12):2422–2433, 2010.

Renato Guseo and Mariangela Guidolin. Market potential dynamics in innovation diffusion: Modelling the synergy between two driving forces. *Technological Forecasting and Social Change*, 78(1):13–24, 2011.

Renato Guseo and Cinzia Mortarino. Sequential market entries and competition modelling in multi-innovation diffusions. *European Journal of Operational Research*, 216(3):658–667, 2012.

Renato Guseo and Cinzia Mortarino. Within-brand and cross-brand word-of-mouth for sequential multi-innovation diffusions. *IMA Journal of Management Mathematics*, 25(3):287–311, 2014.

Renato Guseo and Cinzia Mortarino. Modeling competition between two pharmaceutical drugs using innovation diffusion models. *The Annals of Applied Statistics*, 9(4):2073–2089, 2015.

Renato Guseo, Alessandra Dalla Valle, and Mariangela Guidolin. World Oil Depletion Models: Price effects compared with strategic or technological interventions. *Technological Forecasting and Social Change*, 74(4):452–469, 2007.

Herman O Hartley. Exact confidence regions for the parameters in non-linear regression laws. *Biometrika*, 51(3/4):347–353, 1964.

John Hauser, Gerard J Tellis, and Abbie Griffin. Research on innovation: A review and agenda for marketing science. *Marketing Science*, 25(6):687–717, 2006.

Rob J Hyndman and George Athanasopoulos. *Forecasting: Principles and Practice*. OTexts, 2018.

Michael L Katz and Carl Shapiro. Technology adoption in the presence of network externalities. *Journal of Political Economy*, 94(4):822–841, 1986.

Trichy V Krishnan, Frank M Bass, and V Kumar. Impact of a late entrant on the diffusion of a new product/service. *Journal of Marketing Research*, 37(2):269–278, 2000.

Alfred J Lotka. Analytical note on certain rhythmic relations in organic systems. *Proceedings of the National Academy of Sciences of the United States of America*, 6(7):410–415, 1920.

Vijay Mahajan and Eitan Muller. Innovation diffusion and new product growth models in marketing. *Journal of Marketing*, 43(4):55–68, 1979.

Vijay Mahajan, Eitan Muller, and Frank M Bass. New product diffusion models in marketing: A review and directions for research. *Journal of Marketing*, 54(1):1–26, 1990.

Vijay Mahajan, Eitan Muller, and Yoram Wind. *New-Product Diffusion Models*, volume 11. New York: Springer Science & Business Media, 2000.

Edwin Mansfield. Technical change and the rate of imitation. *Econometrica: Journal of the Econometric Society*, 29(4):741–766, 1961.

Cesare Marchetti. Society as a learning system: Discovery, invention, and innovation cycles revisited. *Technological Forecasting and Social Change*, 18(4):267–282, 1980.

Peter McCullagh and John A Nelder. 1989. *Generalized Linear Models*, volume 37. CRC Press, 1983.

Nigel Meade and Towhidul Islam. Modelling and forecasting the diffusion of innovation–a 25-year review. *International Journal of Forecasting*, 22(3):519–545, 2006.

Geoffrey A Moore. *Crossing the Chasm*. New York: Harper Business, 1999.

Steven A Morris and David Pratt. Analysis of the Lotka-Volterra competition equations as a technological substitution model. *Technological Forecasting and Social Change*, 70(2):103–133, 2003.

Eitan Muller, Renana Peres, and Vijay Mahajan. *Innovation Diffusion and New Product Growth*. Cambridge: Marketing Science Institute, 2009.

Mark Newman. *Networks*. Oxford: Oxford University UPress, 2018.

John A Norton and Frank M Bass. A diffusion theory model of adoption and substitution for successive generations of high-technology products. *Management Science*, 33(9):1069–1086, 1987.

Renana Peres, Eitan Muller, and Vijay Mahajan. Innovation diffusion and new product growth models: A critical review and research directions. *International Journal of Research in Marketing*, 27(2):91–106, 2010.

Fotios Petropoulos, Daniele Apiletti, Vassilios Assimakopoulos, et al. Forecasting: Theory and practice. *International Journal of Forecasting*, 38(3):705–871, 2022.

R Core Team. *R: A Language and Environment for Statistical Computing*. R Foundation for Statistical Computing, Vienna, Austria, 2021. URL https://www.R-project.org/.

Everett M Rogers. *Diffusion of Innovations*. New York: Free Press, 2003.

Nathan Rosenberg. *Perspectives on Technology*. Cambridge: Cambridge University Press, 1976.

Max Roser, Hannah Ritchie, and Esteban Ortiz-Ospina. The internet's history has just begun. *Our World in Data*, 2018. URL https://ourworldindata.org/internet-history-just-begun.

Sergei Savin and Christian Terwiesch. Optimal product launch times in a duopoly: Balancing life-cycle revenues with product cost. *Operations Research*, 53(1):26–47, 2005.

Andrea Savio, Luigi De Giovanni, and Mariangela Guidolin. Modelling energy transition in Germany: An analysis through ordinary differential equations and system dynamics. *Forecasting*, 4(2):438–455, 2022.

Joseph A Schumpeter. The creative response in economic history. *The Journal of Economic History*, 7(2):149–159, 1947.

George AF Seber and Chris J Wild. *Nonlinear Regression*. New York: Wiley, 1989.

Carl Shapiro and Hal R Varian. *Information Rules: A Strategic Guide to the Network Economy*. Harvard Business Press, 1999.

Galit Shmueli. To explain or to predict? *Statistical Science*, 25(3):289–310, 2010.

Pierre-François Verhulst. Notice sur la loi que la population suit dans son accroissement. *Correspondence Mathematique et Physique*, 10:113–126, 1838.

Vito Volterra. Fluctuations in the abundance of a species considered mathematically. *Nature*, 118(2972):558–560, 1926.

Federico Zanghi. *DIMORA: Diffusion Models R Analysis*, 2021. URL https://CRAN.R-project.org/package=DIMORA. R package version 0.2.0.

Index

Innovation Diffusion Models: Theory and Practice, First Edition. Mariangela Guidolin.
© 2024 John Wiley & Sons Ltd. Published 2024 by John Wiley & Sons Ltd.
Companion Website: www.wiley.com/go/innovationdiffusionmodels/